The Quarters and the Fields

NEW PERSPECTIVES ON THE HISTORY OF THE SOUTH

UNIVERSITY PRESS OF FLORIDA

Florida A&M University, Tallahassee
Florida Atlantic University, Boca Raton
Florida Gulf Coast University, Ft. Myers
Florida International University, Miami
Florida State University, Tallahassee
New College of Florida, Sarasota
University of Central Florida, Orlando
University of Florida, Gainesville
University of North Florida, Jacksonville
University of South Florida, Tampa
University of West Florida, Pensacola

The Quarters and the Fields

Slave Families in the Non-Cotton South

Damian Alan Pargas

UNIVERSITY PRESS OF FLORIDA
Gainesville · Tallahassee · Tampa · Boca Raton
Pensacola · Orlando · Miami · Jacksonville · Ft. Myers · Sarasota

Copyright 2010 by Damian Alan Pargas
Printed in the United States of America on acid-free paper.
All rights reserved

First cloth printing, 2010
First paperback printing, 2011

Library of Congress Cataloging-in-Publication Data
Pargas, Damian Alan.
The quarters and the fields : slave families in the non-cotton South / Damian Alan Pargas.
p. cm.—(New perspectives on the history of the South)
Includes bibliographical references and index.
ISBN 978-0-8130-3514-7 (alk. paper)
ISBN 978-0-8130-3804-9 (pbk.)
1. Slaves—Family relationships—Southern States. 2. Slaves—Southern States—Social conditions. 3. Plantation life—Southern States—History. 4. Slavery—Southern States—History. 5. African American families—Southern States—Social conditions—History. 6. Agriculture—Social aspects—Southern States—History. 7. Geography—Social aspects—Southern States—History. 8. Southern States—Social conditions.
I. Title. E443.P37 2010
975.'041—dc22   2010017081

The University Press of Florida is the scholarly publishing agency for the State University System of Florida, comprising Florida A&M University, Florida Atlantic University, Florida Gulf Coast University, Florida International University, Florida State University, New College of Florida, University of Central Florida, University of Florida, University of North Florida, University of South Florida, and University of West Florida.

University Press of Florida
15 Northwest 15th Street
Gainesville, FL 32611-2079
http://www.upf.com

Contents

List of Tables   vii
Acknowledgments   ix

PART I. RETHINKING THE EXPERIENCES OF SLAVE FAMILIES

Introduction: Agency, Diversity, and Slave Families   3
1. Three Slave Societies of the Non-Cotton South   13

PART II. THE BALANCING ACT: WORK AND FAMILIES

2. The Nature of Agricultural Labor   39
3. Family Contact during Working Hours   63
4. Family-Based Internal Economies   88

PART III. SOCIAL LANDSCAPES: FAMILY STRUCTURE AND STABILITY

5. Slaveholding across Time and Space   117
6. Marriage Strategies and Family Formation   142
7. Forced Separation   171

PART IV. CONCLUSIONS

8. Weathering Different Storms   203

Notes   207
Bibliography   235
Index   249

Tables

Table 1.1. Selected Agricultural Production in Fairfax County, 1850 and 1860 18
Table 1.2. Population of Fairfax County, Virginia, 1800–1860 20
Table 1.3. Rice Production in Georgetown District, South Carolina and the United States, 1840–1860 26
Table 1.4. Population of Georgetown District, South Carolina, 1800–1860 27
Table 1.5. Population of St. James Parish, Louisiana, 1810–1860 35
Table 1.6. Annual Sugar Production (in hogsheads) in St. James Parish and Louisiana, 1849–1860 35
Table 5.1. Distribution of the Fairfax County Slave Population by Slaveholding Size, 1810–1860 118
Table 5.2. Distribution of the Georgetown District Slave Population by Slaveholding Size, 1800–1860 124
Table 5.3. Distribution of the St. James Parish Slave Population by Slaveholding Size, 1810–1860 130

Acknowledgments

Researching and writing this book took me, all told, about five years. Along the way—as I developed a vague idea about American slave families and turned it into a book—I was fortunate enough to have ample support from family, colleagues, friends, and institutions. Without their help this book would never have gotten off the ground.

I extend my thanks first and foremost to my family members in all three geographic clusters of the Atlantic world. My family in the United States—Fernando Pargas, Denise and Charlie Errico, Gabriel Pargas, Alexandra Pargas, Elizabeth (Nanny) McDermid—provided me with invaluable advice, encouragement, and luxurious accommodation during my research trips to northern Virginia and Washington. My family in Uruguay—Héctor and Ester (Tata & Abuelita) Pargas, Diana Pargas, Mariana Gesto, Daniela Gesto, Néstor Pargas, and little Lucia Pargas—has always been a lasting source of emotional support and generously provided me with a sunny and beautiful destination for several much-needed vacations. Finally, my family in the Netherlands—Peter Bos, Janny Bos, Maya Bos, and Marloes Bos—provided me with encouragement and support, without which I would have been unfit to perform my duties.

At Leiden University Chris Quispel and Piet Emmer in particular believed in me and in my ideas long before I did. Both offered invaluable advice. For their support and wisdom I am infinitely grateful. My other colleagues at Leiden University also provided me with feedback, advice, coffee, and every now and again a free lunch, especially Filipa Ribeiro da Silva, Chris Nierstrasz, Jessica Roitman, Cátia Antunes, Hans Wilbrink, Jorrit van den Berk, Andreas Weber, Alicia Schrikker, Job Weststrate, Marijke Wissen-van Standen, Leonard Blussé, Henk Kern, Joost Augusteijn, Leo Lucassen, Peter Meel, Gert Oostindie, Eduard van der Bilt, and Adam Fairclough.

There are many others, too numerous to mention here, who also deserve my sincere thanks. My colleagues at *Itinerario*—Frans-Paul van der Putten, Alicia Schrikker, Annelieke Dirks, Gijs Kruijtzer, and Lincoln Paine—also provided me with invaluable editing experience and a welcome diversion from writing. I extend my warm thanks across the ocean to Professor Stanley Engerman at the University of Rochester for his advice and cooperation. My closest colleagues at Utrecht University—especially Jaap Verheul, Derek Rubin, and Rob Kroes—provided me with a pleasant environment in which to finish this book. Finally, my friends Alexander Hoorn and Johan Kwantes forced me on numerous occasions to get out of the office and relax for the sake of my own sanity.

Several institutions provided me with financial assistance as well as helpful feedback on this manuscript, none more so than Leiden University, which funded most of my research trips to the United States, paid me a generous salary, and provided me with a pleasant workplace. The Leids Universitair Fonds (LUF) financed my research trip to Louisiana in the summer of 2006. The N. W. Posthumus Institute for Social and Economic History provided me with an excellent opportunity to present my research at various stages and offered me constructive criticism, most of which I have applied to the obvious improvement of this manuscript. Special thanks to Ben Gales for his advice and help. The Wissenschaftskolleg zu Berlin also provided me with a luxurious forum at which to present my research in the winter of 2006 and offered constructive criticism. The numerous institutions I visited in the United States gave me pleasant workplaces away from home, and their librarians and assistants were quite helpful and accommodating. Especially the staff at the Library of Congress and National Archives in Washington and at Louisiana State University's Hill Memorial Library were particularly friendly and well informed, and they deserve special mention here. The editors and peer reviewers at the *Journal of Family History* and *American Nineteenth Century History* provided me with excellent comments and tips, as well as a platform for presenting earlier versions of some of the chapters in this book. Finally, the staff and editors at University Press of Florida have been extremely friendly, helpful, and accommodating during all stages of turning this manuscript into a book. I especially extend my warm thanks to Meredith Babb and Heather Turci; also to John David Smith (the series editor), Patterson Lamb (for the excellent copyediting), and the anonymous peer reviewers of my original manuscript.

Most of all I thank my loving wife, Tamara, who has been a source of inspiration and encouragement to me during the past decade, who accompanied me during several research trips, who edited portions of this manuscript, and to whom I dedicate the finished product.

I

Rethinking the Experiences of Slave Families

Introduction

Agency, Diversity, and Slave Families

Almost a century and a half have passed since the fiery collapse of slavery and the emancipation of over four million African Americans held in bondage in the American South. In recent years a vast outpouring of research has rightfully salvaged slavery from the margins of American history and thrust it into the spotlight; yet despite the publication of hundreds of books and articles on the subject, our understanding of many aspects of enslaved people's social lives remains clouded by disagreement among contemporary scholars. The nature of slave family life has proved to be an especially thorny issue, and a general consensus among historians regarding the daily experiences, structure, and stability of families in bondage has been slow in coming.

A survey of the historical literature suggests that two specific issues lie at the root of this disagreement. First, scholars have long disagreed over the extent to which slave family life was shaped by either external forces (the economy, slaveholders, the law) or slave agency (the actions of enslaved people themselves). Most historians have tended to emphasize one view to the exclusion of the other. Second, scholars of antebellum slavery in particular (roughly the period 1800–1860) have long underestimated geographic differences among slave families and continue to disagree over which characteristics of slave family life can be considered "typical" for the South as a whole. Indeed, despite the acknowledgment of a traditional overemphasis on slave culture in the cotton South—which led to a wave of regional studies among the past generation of scholars, many of which illuminate slavery in various marginal communities of the non-cotton South—many recent studies still draw very generalized conclusions from localized research, and few have employed an intraregional comparative approach. In short, many studies have tended to take an exclusive, rather than inclusive, approach to slave culture and slave families. By emphasizing agency to the exclusion of external factors, for example, or presenting family life in one region as "typical" for the entire South, historians have often underestimated the dynamics and diversity of both slave family life and the antebellum South.[1]

Focusing on the experiences of slave families in the non-cotton South, this book provides a reinterpretation of enslaved people's family lives, namely, by formulating a middle ground in the historical debate over slave agency and by redefining slave family life in plural form. First, this work argues that the varied nature of regional agriculture in diverse southern localities was the most important underlying factor in the development of slave family life—not because it dictated the experiences of slave families from above per se, but because it confronted them with a basic framework of *boundaries and opportunities* with respect to family contact, child care, family-based internal production, marriage strategies, and long-term stability. Second, this book underscores the diversity of slave family life in different agricultural regions of the nineteenth-century South. A comparative study that examines the importance of time and place for slave families, it aims to advance a pluralist view of families' experiences in bondage, positing that regional differences between slave families were the rule rather than the exception.

Specifically, this book examines how the nature of regional agriculture affected "simple" slave families (consisting of couples, whether co-residential or not, with or without children, or singles living with their children) in three very different parts of the non-cotton South: northern Virginia, lowcountry South Carolina, and southern Louisiana. In the following, this study will be placed within a broader historiographical context and its approach and methodology further explained.[2]

Boundaries and Opportunities: The Extent of Agency

One of the aims of this book is to suggest a new way of thinking about the extent of agency in shaping slave culture and especially family life. As such, it builds upon more than a century of scholarship, during much of which a top-down perspective of slavery—and at best only a scant interest in slave culture and family life—prevailed. Prior to the 1970s scholars generally viewed the slave family as a catastrophic failure, and historians tended to attribute enslaved people virtually no agency (or indeed even interest) in trying and establishing anything approaching stable or cohesive families.

Explanations ranged from the racist (roughly until World War II) to the defeatist (in the 1950s and 1960s). For example, when the works of early-twentieth-century white scholars such as U. B. Phillips—whose *American Negro Slavery* (1918) set the standard for decades—broached the subject of slave culture or family life at all, they tended to confirm prevailing racist stereotypes, implying that character deficiencies among African Americans rendered them incapable of developing morally sound family values. Historians depicted licentious and irresponsible bondspeople whom paternalistic slaveholders desperately strove to civilize by encouraging marital unions, punishing adultery, establishing rules for the care and supervision of slave children, and protecting pregnant women from overwork. Yet according to these scholars, the efforts of benevolent masters were often in vain, as African Americans' perceived childlike nature still made slave families loosely organized and unstable.[3]

Causal explanations for the "failure" of slave families changed fundamentally in the 1950s, however, as the focus of scholars shifted from the perceived character deficiencies of African Americans to the dehumanizing brutality of the peculiar institution itself. Like African-American sociologists W. E. B. Du Bois and E. Franklin Frazier before them, postwar historians such as Kenneth Stampp indicted American slavery and suggested that slave family life had in every possible way been truncated by the oppressive institution of human bondage. The new consensus that emerged during this period was that slave families had been loosely organized because they had lacked legal standing and had been the victims of abuse, overwork, rape, white interference, and forced separation. Moreover, the trauma of slavery had supposedly rendered African Americans virtually incapable of maintaining monogamous and stable marriages, and psychologically incapacitated them for the adult responsibilities of child rearing.[4]

A common denominator in the research that supported the dominant paradigm of the slave family as having been either encouraged and protected, or abused and truncated, by slaveholders was that it interpreted slave culture and family life almost exclusively from a top-down perspective. Historians generally accepted contemporary whites' perceptions of black families; they ignored sources left by slaves themselves and failed to analyze slave family life from the perspective of enslaved people. Referring to the works of the 1950s and early 1960s, historian John Blassingame lamented in the 1970s that the institution of slavery had erroneously been depicted as a "monolithic institution which [stripped] the slave of any meaningful and distinctive culture, family life, religion, or manhood." According to Blassingame and many others, scholars had concentrated too much on *what the institution of slavery did to men and women in bondage*, as if enslaved people had been passive victims for over two and a half centuries. This paradigm of slave culture came increasingly under attack in the late 1960s and early 1970s, a period that also witnessed the emerging contours of a debate between revisionist historians of slavery on the one hand, and cliometricians on the other.[5]

The notion that enslaved people had been passive victims of a totalitarian system that stripped them of their humanity was rejected outright by revisionists such as Eugene Genovese, Herbert Gutman, and John Blassingame, among others.[6] Instead of concentrating on what the institution of slavery did to enslaved people, revisionist studies underscored slave agency, or *what enslaved people did for themselves during bondage*. Most consulted previously untapped sources in order to shift the focus from the planters to the slaves, including slave autobiographies, antebellum interviews with fugitive slaves in the northern states and Canada, and interviews with elderly ex-slaves conducted in the 1930s by workers of the Federal Writers' Project of the Works Progress Administration (WPA).[7] Reinterpreting the lives of enslaved people, writers of revisionist studies celebrated an autonomous slave culture that flourished despite the horrors of bondage. They concluded that enslaved people resiliently weathered the storm of slavery by seeking comfort in one another, establishing stable, monogamous relationships and a loving and cohesive family life. Revisionists emphatically attributed the establishment and maintenance of family relationships to the remarkable actions of the enslaved, not to any encouragement or protection by slaveholders. Gutman, in his classic study *The Black Family in Slavery and Freedom* (1976), even argued that slaves' dedication to family resulted in long-lasting marriages and two-parent households throughout the South, contrary to what historians previously had believed.[8]

While revisionist historians celebrated the success of the slave family as a remarkable testament to slave agency, cliometricians Robert Fogel and Stanley Engerman simultaneously presented new arguments that came to similar conclusions but retained an emphasis on the external forces that shaped slave family life. In their statistical analysis of American slavery, *Time on the Cross* (1974), Fogel and Engerman argued that slave families were indeed stable, though not so much because of any resilience on the part of enslaved people but rather because the economics of slavery were conducive to family stability. According to Fogel and Engerman, the establishment of slave families was encouraged and protected by capitalist slaveholders, whose actions stemmed not from benevolence (as Phillips believed) but primarily from economic interests—family stability, after all, produced better workers, discouraged flight and rebellion, facilitated labor organization, and simplified the distribution of rations and housing to the labor force. Fogel and Engerman theorized that the capitalist nature of slavery not only encouraged slave family formation but also protected them from, among other things, forced separation and sexual interference. The belief that family life had been truncated and obstructed by southern masters was rejected by cliometricians as an abolitionist myth, and the role of slave agency in shaping family life was significantly downplayed.[9]

Both camps of this debate had their shortcomings. Revisionists focused so much of their attention on slave agency and enslaved people's autonomous culture that they lost sight of the broader framework of slavery, especially the limitations it imposed on agency and family life. Historian Peter Parish, among others, criticized the revisionist tendency to write "the slaveholders out of the story completely."[10] The cliometric argument that slave families were little more than protected agents of a capitalist system, however, went to the opposite extreme by downplaying the efforts of enslaved people to establish and protect their families under admittedly difficult circumstances, attributing too much goodwill to slaveholders and underestimating some of the major obstacles that slave families certainly encountered.[11]

Were slave families stable and cohesive during slavery, then? And can the successes or failures of slave family life be attributed to the actions of enslaved people themselves or to external forces such as the nature of the southern economy and the intervention of slaveholders? These questions continue to fuel debates within the academic community and remain as yet unresolved. Despite widespread criticism of the revisionist exaggeration of slave autonomy, recent studies of slave culture and family life have more often than not tended to retain an overemphasis on slave agency, and the past few years have even witnessed what appears to be the beginning of a backlash by historians such as Wilma Dunaway, whose rejection of the "academic exaggeration of slave agency" is scathing and who has shifted the focus back to the disastrous effects of external factors on slave families. Scholars' understanding of what the institution of slavery did to families in bondage need not cancel out our understanding of what slave families did for themselves, however, and upon closer inspection the polarized camps of this historical debate are not as mutually exclusive as they may seem. Most scholars now agree that external factors *and* slave agency were crucial determinants in the family lives of the enslaved—the trick lies in determining the proper weight to assign to each.[12]

With the intention of presenting a broader view of the complex forces that shaped enslaved people's family lives, not only from outside but also from within, this book takes an inclusive (rather than exclusive) approach to the slave agency debate. Here the concept "boundaries and opportunities" will be advanced as a useful analytical framework for understanding the reality of slave family life. The concept underscores not only force but also the importance of *intent* and *possibility* in the actions and/or inaction of enslaved people.

There can be no doubt that external factors limited and often eliminated choices for bondspeople. Slaves' attempts to maximize family contact, establish what they perceived to be acceptable domestic arrangements, and achieve long-term stability were all circumscribed and frequently frustrated by the legal, social, and economic boundaries of slavery. Slave families were not established in an autonomous vacuum and the success of slave agency in shaping a cohesive family life should not be exaggerated. Yet neither should the efforts of enslaved people in shaping their family lives be dismissed or ignored, because when external factors provided families with limited opportunities to achieve more stability, more time and more autonomy, however minor or circumscribed, most actively seized them. Indeed, some tested the boundaries and attempted to create new opportunities for family life, even when the chances for success were limited. Agency should not be confused with success. Any attempt on the part of an enslaved person to establish a family or make family life more attractive demonstrated agency, even if their attempt failed due to external factors. Slave families were thus shaped by boundaries and opportunities, external forces and internal forces, however imbalanced. The former did not always truncate family life—nor did they always protect families—and the latter certainly did not always triumph.

In the present study the external forces that were directly related to the nature of regional agriculture form the key area of focus, as they were of particular importance in laying the foundations for slave family life and were largely responsible for its diversity across time and space. Rather than concentrate exclusively on what these factors may or may not have done to enslaved families directly, however, this study aims to determine what these factors meant for enslaved families in practice—what kinds of *boundaries* they imposed and what kinds of *opportunities* they created—without losing sight of how enslaved people as historical actors reacted to those boundaries and seized or created opportunities to shape their family lives. By illuminating slaves' *intentions* with respect to their family lives, as well as the absence or availability of *possibilities* to actually realize their ideals, this study aims to avoid an overemphasis on agency and formulate a middle ground in the traditional slave agency debate. In order to analyze slave families from both internal and external perspectives the source material consulted for this book runs the full gamut of available evidence left by masters (including memoirs and plantation records), slaves (especially interviews and slave narratives), and third parties (such as travelers' accounts and government data).

A Comparative Regional Study: Underscoring Diversity

A second aim of this book is to underscore the geographic diversity of slave family life. Throughout the twentieth century, scholars tended to paint a singular and homogenized picture of the South, both of the institution of slavery and of the experiences of slaves. Peter Parish opined in 1989 that "in attempting to treat the subject at large, Stampp, Elkins, Blassingame, Genovese, Fogel and Engerman, and several others, all tend to flatten out differences and variations . . . and to pay inadequate attention to slavery in its more unusual forms." Philip Morgan later lamented in the introduction of his comparative study of eighteenth-century slave culture, *Slave Counterpoint* (1998), "Too often in history one South has served as proxy for many Souths." With respect to the nineteenth century, historians traditionally focused on the experiences of slaves living in the cotton districts. In doing so they underestimated the social and economic diversity of the antebellum South and presented a virtually homogenous picture of enslaved people's family lives.[13]

The past generation of historians has successfully avoided an overemphasis on the cotton South and produced a number of fine regional studies on slave families in the non-cotton districts. As yet, however, they have failed to produce many works that transcend localized research and employ a broader comparative approach. Larry Hudson produced an excellent study of slave family life in different geographic and economic sections of South Carolina, *To Have and to Hold* (1997), but differences within that state were not as dramatic as they were between different geographic regions of the South. The same goes for Ann Patton Malone's classic study of slave families in the cotton and sugar districts of Louisiana, *Sweet Chariot* (1992); Emily West's *Chains of Love* (2004), which also deals with South Carolina;

and more recently Daina Ramey Berry's *Swing the Sickle for the Harvest Is Ripe* (2007), which focuses on Georgia. These are all valuable works that take into account regional differences within certain states, but a broader comparative study of slave families in different states would expand our understanding of the diversity of slave family life in the antebellum South as a whole. It would also be less susceptible to the kinds of generalizations that have characterized some regional studies.[14]

This book intends to fill that gap by comparing and contrasting the experiences of slave families living in three small counties located in very different sections of the antebellum South: Fairfax County in northern Virginia (the Chesapeake), where the staple crops were wheat, corn, and other grains; Georgetown District in the rice-producing coastal region of South Carolina (the lowcountry); and St. James Parish, located in the heart of the southern Louisiana sugar country (the Lower Mississippi Valley).[15] By focusing on counties, this study intends to magnify slave family life in three communities that typify the regions in which they were located. (Sources from neighboring counties in each region have also been consulted, both for illustrative purposes and to place the counties within a regional context.) None of these antebellum communities or regions lay in the so-called Cotton Kingdom, but nevertheless they all provide excellent case studies for slave family life because they perfectly illustrate the variations in the boundaries and opportunities with which families living in different agricultural regions were confronted. Not only did these slave families live in different localities and cultivate different staple crops, but social landscapes and long-term economic developments in each of the three counties also differed widely.

Indeed, the regions chosen for this study represent the extremes of the southern economy and southern slavery. Fairfax County was a devolving slave society in the nineteenth century, characterized by severe economic decline in slave-based agriculture, a diminishing slave population, and work patterns that reflected the nature of mixed farming instead of monoculture plantation agriculture. By contrast, Georgetown District was a more stable and extremely affluent slave society, characterized by large slaveholdings and flexible labor arrangements that provided enslaved people with a number of unique opportunities to shape their lives in bondage. And St. James Parish showed all of the telltale signs of a rapidly expanding slave society at the epicenter of a booming sugar industry—slaveholding size underwent mushroom growth (skewed toward a male majority) and work patterns combined the precision of military drill with factory-like shifts. The structural differences in all of these regions had widely divergent consequences for the experiences of slave families.

By incorporating the regional study with the comparative approach, this work aims to tease out the different lives of enslaved families living in different geographic localities, especially as these differences related to the nature of regional agriculture. As a result, this book will demonstrate that slave families in different agricultural regions were confronted with different boundaries and opportunities, and that family life was thus very much a plural phenomenon in the antebellum South.

Themes and Chapter Synopsis

As stated earlier, this study examines how the nature of regional agriculture affected slave families living in various localities of the non-cotton South. The topic will be tackled thematically, with chapters focusing on the daily experiences, structure, and stability of slave families across time and space. Two broad themes in particular stand out, however, which form the basic outline of this book: work and social landscapes.

After the first chapter, which will introduce the reader to the three slave societies chosen for this study and provide a broad overview of the slave-based economies around which these rural communities revolved, part II delves into the boundaries and opportunities created by work for families in bondage in northern Virginia, lowcountry South Carolina, and southern Louisiana, respectively. Chapter 2 lays the groundwork by examining and explaining the daily and seasonal tasks and work patterns of enslaved people in each of the three regions. The focus of chapter 3 is on slave families themselves, specifically on the boundaries and opportunities that work and regional agriculture created for family contact during working hours. Particular attention is paid to families' experiences with pregnancy and child care, and the extent to which slave family members were afforded opportunities to work together in the fields. And chapter 4 traces the development of slave families' internal economies to their formal work and the nature of regional agriculture in the regions where they lived. Especially the consequences of work and agriculture for the time and physical means that families had to improve their own material conditions are examined.

In part III the effects of regional agriculture on enslaved people's social landscapes are explored, with a particular emphasis on family structure and stability across time and space. Chapter 5 analyzes the effects of regional agriculture on the development of slaveholding size, as well as the spatial distribution and sex ratios of enslaved populations in each region during the antebellum period. The consequences of various demographic boundaries and opportunities for slaves' marriage strategies and family structure are the subject of chapter 6. And in chapter 7 the long-term stability of families in bondage is analyzed by examining their experiences with forced separation—specifically through sale, estate divisions, and long-term hiring. Finally, the conclusions of this study are clarified in the final chapter.

It is not the intention of this study to provide a definitive history of slave family life in each of the three regions chosen—each of which merits an entire volume unto itself—nor of the non-cotton South in its entirety. Moreover, factors such as religion or African cultural continuities will not be discussed, both of which were of undeniable importance to slave families. Rather, the aim of this book is simply to suggest a new way of thinking about antebellum slave families, to trace the foundations of slave family life to the rural economies in the regions to which they were bound, to explore the boundaries and opportunities created by various slave-based economies, and to illuminate the importance of time and place for slave family life.

1

Three Slave Societies of the Non-Cotton South

What kinds of staple crops dominated slave-based agriculture in northern Virginia, lowcountry South Carolina, and southern Louisiana? To what extent was the cultivation of these crops profitable? Did the various communities in which slave families lived experience economic decline, stability, or rapid growth in the antebellum period? The answers to these simple queries are crucial for rethinking our understanding of slave families' experiences. Throughout the rural South it was the land upon which enslaved people worked, and which had been worked by their forebears, that most strongly influenced the terms of their bondage and the nature of their social lives. A general discussion of the evolution of slave-based economies in the three regions chosen for this study is therefore a necessary starting point for examining the effects of regional agriculture on slave families.

When analyzing the trajectory and rate of development of these economies, it is useful to apply the concepts "slave societies" and "societies with slaves." Virtually every region of the South constituted at some point in its history a thoroughly entrenched slave society, characterized by an agricultural sector in which, in the words of Ira Berlin, "slavery stood at the center of economic production, and the master-slave relationship provided the model for all social relations." In societies with slaves, by contrast, slave labor was marginal to the local economy, constituting but "one form of labor among many." Economies are dynamic by nature, however, and in the nineteenth century not all regions were developing in the same direction or at the same rate. Different crops, cultivated under different circumstances, were susceptible to different trends and different degrees of success within the broader southern economy. These variations in turn resulted in different kinds of boundaries and opportunities for slave families, as will become clear throughout this book.[1]

This chapter outlines the nature of slave-based agriculture in nineteenth-century Fairfax County, Virginia; Georgetown District, South Carolina; and St. James Parish, Louisiana, respectively. Providing a brief overview of the introduction and development of the staple crops cultivated in each of the three regions, this chapter sets the stage for a more in-depth study of the relationship between regional agriculture and slave family life by exploring the workings of the local economies upon which their fates depended.

Down (and Out) on the Farm: Fairfax County, Virginia

Few slaveholding regions in nineteenth-century America struggled to stay afloat as much as did the Upper South, and contemporary discussions concerning the future of bondage in local agriculture there were largely characterized by widespread pessimism. In the rejected draft of an address to the Virginia State Agricultural Society in 1852, for example, one disillusioned farmer spoke for his class when he lamented that while the "southern states stand foremost in agricultural labor," he and his fellow Virginians should find "no cause for self-gratulation." Typical of the statewide trend—but atypical for the South as a whole—slave-based agriculture in Fairfax County suffered significant decline during the antebellum period, causing a once thriving tobacco-based slave society to devolve steadily in the direction of a wheat-based society with slaves.[2]

Initial developments in colonial northern Virginia certainly seemed to point toward a promising future, however. Bordering the Potomac River to the north and southeast, Fairfax County emerged in the early eighteenth century as a tidewater plantation society, both culturally and economically bound to the thriving Chesapeake region that encompassed the easternmost sections of Virginia and Maryland. In the Chesapeake, tobacco was king. The source of spectacular wealth for many enterprising planters—both in Fairfax and elsewhere along the tributaries of the Chesapeake Bay—the weed was planted as quickly as the forests could be cleared. "Like water for irrigation in a dry country," as historian Frederick Gutheim put it, "tobacco alone made land [along the Potomac] valuable."[3]

Following the lead of their counterparts in other tidewater counties, Fairfax planters quickly turned to the slave trade to meet their insatiable demand for labor; between 1732 and 1772 dozens of merchant vessels arriving from the "Coast of Africa," Barbados, Antigua, and other slave-trading hubs, delivered large groups of Africans to toil on the expanding tobacco plantations. By the time Fairfax officially received county status in 1742, approximately 29 percent of its population already consisted of men and women in bondage. As the century neared its close, the enslaved population had grown to 41 percent. Fairfax rapidly developed into a slave society.[4]

The triumph of tobacco proved short-lived, however. Throughout the Chesapeake, and especially in northern counties such as Fairfax, the last quarter of the eighteenth century witnessed a dramatic shift away from the crop that had once been so vital to the region; most planters would hastily abandon it within the span of one generation. The major reason for this shift was soil exhaustion, which consistently plagued local plantation agriculture with decreased productivity and diminished quality. Tobacco cultivation exhausted the once fertile soil of Fairfax County to a point of such sterility that by the second half of the eighteenth century an impending economic crisis seemed poised to seal the fate of local planters.[5]

Virginia tobacco planters were dealt an additional blow when the Revolutionary War broke out, which, according to economic historian Avery Odell Craven, "acted as a powerful force in bringing disaster to much that was already on its way to ruin, and gave added impulse to many of the changes already begun." During the conflict, direct export to Great Britain, the colonies' most important commercial market, was cut off. Reaching other overseas markets such as France and Holland proved challenging as well—Atlantic shipments were subject to confiscation by the British navy, and blockades in the Chesapeake Bay prevented or delayed departures, to the acute frustration of Virginia tobacco planters and merchants. All in all, tobacco exports declined substantially during the war. The total exports for the entire period between 1776 and 1782 were less than the exports of a single year just preceding the outbreak of hostilities, and Thomas Jefferson estimated in 1787 that no less than two-thirds of the tobacco shipped from Virginia during the war had been captured by the British. Prices had taken a nosedive, as tobacco lost as much as 80 percent of its value.[6]

Soil exhaustion and decreased productivity, combined with the disastrous effects of the war with Great Britain, caused tidewater tobacco planters to lose faith in their traditional cash crop by the end of the eighteenth century. The lavish standard of living once enjoyed by the Fairfax elite became difficult, even impossible, to maintain, and indeed the children and grandchildren of colonial tobacco barons found themselves increasingly hard put to turn a profit at all. Some decided to cut their losses and manumit their slaves. Others chose to leave Fairfax and Virginia behind, determined to try their luck in the expanding and promising western frontier. When Daniel McCarty the Younger inherited his grandfather's tobacco plantation Mount

Air in the late 1790s, for example, he found the soil so sterile and worthless, and the tobacco industry so badly damaged, that he felt compelled to sell his lands with the intention of emigrating from Virginia. He died before he could find a buyer, but some of his peers simply abandoned their estates, which were quickly reclaimed by the forests, and moved south and west of the Appalachians. The turn of the century witnessed the first trickles of what would later become a wave of emigration, a development that would eventually result in significant population decline during the antebellum period.[7]

Those who stayed in Fairfax found themselves in desperate need of spreading their financial risks and redirecting their agricultural pursuits to meet the demands of the current market. Around the turn of the nineteenth century, local planters increasingly intensified their cultivation of grains and a number of other foodstuffs, which had been introduced as secondary staples during the tobacco era. In the minds of many planters, agricultural diversification seemed to offer the only viable way out of what would have otherwise been certain bankruptcy. Domestic and foreign demand for foodstuffs increased both during and after the Revolutionary War, and prices for wheat and corn almost doubled in the closing decades of the eighteenth century. The situation was given an added impulse when upon the outbreak of the French Revolution in 1789 an acute scarcity of grain developed throughout Europe as well as the West Indies. Fairfax County planters therefore began to shift their energy and attention to the commercial production of wheat and corn—later supplemented by other small grains such as oats and rye—for both domestic and foreign markets.[8]

As the production of wheat and corn increased, tobacco cultivation declined until it was virtually eliminated from the county. This shift was not undertaken by all Fairfax County planters at the same time (George Washington had begun as early as the 1760s), but ultimately they all followed the same general trend. According to one local historian, "by the early 1800's the farmers were no longer depending on tobacco, and in fact, there is little evidence of much tobacco being grown here at that time." Elijah Fletcher, a young tutor from Vermont who was briefly employed to teach the children of Thomson Mason at Hollin Hall plantation, wrote to his father in 1810 that "the staple commodity or most general crop [here] is wheat and corn. They have abandoned the cultivation of Tobacco in a great degree, it requiring a very rich soil and much attention, they do not find it profitable."[9]

While wheat, corn, and other small grains became the primary staple crops during the antebellum period, tobacco production did not disappear completely from Fairfax County. Especially during the early decades of the nineteenth century it was still cultivated by some farmers on a small scale to supplement their income from grains. John Davis, an English visitor to Fairfax County at the turn of the nineteenth century, recorded that Pohoke plantation, where he spent a few days, produced mostly "Indian corn, wheat, rye and tobacco." Richard Marshall Scott, Sr., the owner of Bush Hill and Dipple plantations, felt compelled to haul his tobacco—which was stored in an Alexandria warehouse awaiting shipment—back to his estate during the British invasion of Washington in the War of 1812—officially "to keep it out of reach of the enemy, who out of 104 hogsheads only took one." Evidence suggests that at least a few other local planters also cultivated small amounts of tobacco. As late as the 1820s, advertisements for tobacco could still be found in local newspapers, such as the following: "TOBACCO—The subscriber will buy and sell for a commission of one dollar per hhd. Planters and others who have tobacco for sale . . . will have the same promptly attended to." However, with time such advertisements became increasingly scarce and were moreover heavily outnumbered by advertisements calling for wheat and other grains like the following: "Wheat Wanted: The subscriber . . . respectfully solicits a call from farmers and others who have wheat for sale, and will at all times give the Alexandria market price for good wheat"; and "Wheat, Rye and Corn: Lindsay, Hill & Co. continue to purchase wheat, rye and corn."[10]

Table 1.1 illustrates just how negligible tobacco had become by the decade preceding the Civil War. In the agricultural census of 1850 it was omitted altogether, while in 1860 its production was limited to only 29,190 pounds. Corn and oats were cultivated in especially large numbers as they served not only as staple crops but also as food for the planters, slaves, and livestock. Wheat, on the other hand, was grown almost exclusively for sale. Its relatively meager production during the 1850s had to do with the ravishing effects of the jointworm, a pest that threatened to compel local farmers to abandon wheat in favor of corn during that time. Farmers also raised market produce such as potatoes, meat, and dairy products and produced substantial amounts of wool for sale in the nearby urban centers of Alexandria and Washington. One local farmer advised his neighbors to "keep a dairy and cut hay, as butter and hay pay a better profit than grain and stock." As a rule, several products were found to be more reliable than one, and farmers concentrated on producing as many different commodities as they could.[11]

Table 1.1. Selected Agricultural Production in Fairfax County, 1850 and 1860

	1850	1860
Wheat (bushels)	56,150	59,318
Corn (bushels)	207,531	263,225
Oats (bushels)	76,798	155,409
Rye (bushels)	5,860	15,155
Tobacco (pounds)	—	29,190
Irish and Sweet Potatoes (bushels)	28,181	56,171
Butter and cheese (pounds)	144,872	166,676
Wool (in pounds)	16,502	14,391
Value of animals slaughtered (in dollars)	80,452	68,490

Source: U.S. Nonpopulation Census, Agriculture, 1850–1860, National Archives and Records Administration.

The new staple crops did not, however, make many planters rich. The prominent local slaveholder George W. P. Custis, master of Arlington plantation, voiced a common sentiment when in 1827 he admitted to agriculturalist Samuel Janney: "I am accounted one of the richest men in Virginia, yet I seldom have a dollar." Especially foreign markets for grains were susceptible to unpredictable fluctuations and were thus not always reliable. The high demand for wheat and grains in Europe following the French Revolution, for example, initiated a boom in local exports, but this trade was based almost exclusively on a wartime situation. Predictably, it lasted only until the end of the Napoleonic wars. Back home, the War of 1812 also damaged the local grain industry. In 1813 and 1814, Fairfax County wheat shipments to Europe and the West Indies became stranded in Alexandria when the British imposed a blockade of the Chesapeake Bay. Moreover, the strict revision of the Corn Laws in 1815 effectively cut off British markets until their repeal in 1846. American exports of grain and flour to Great Britain plummeted from almost three million barrels in 1801 to none at all in 1815. Although the domestic market for grain remained relatively steady, and exports picked up again by the 1830s (especially to the Iberian Peninsula), most northern Virginia planters struggled throughout the first half of the nineteenth century to turn a profit.[12]

In Fairfax County the fluctuating demand for grain was further exacerbated by continued low crop yields due to soil depletion and the failure to employ advanced farming methods until the 1840s. Local farmers generally plowed their estates with no regard for land contours, which led to massive soil erosion. They also neglected to fertilize their fields properly. Relatively low crop yields provided planters with little capital to invest and thus few opportunities for expansion in either land or slaves, the latter of which became a financial burden and were increasingly gotten rid of. Agricultural operations dwindled with time. By the 1840s much of the available land in Fairfax County was no longer even under cultivation after generations of planters and slaveholders had abandoned their barren farms and emigrated westward. Richard Marshall Scott of Bush Hill recorded a common sight when he wrote in his diary on 4 October 1820: "My friend and brother-in-law, Charles J. Love, moved from Clermont [plantation] with his whole family this day . . . to go to Tennessee to settle on a farm."[13]

In the 1840s, however, Quaker immigrants from the northern states began to pour into the county and reverse the downward spiral. Attracted by cheap land and willing to invest the time and energy to revitalize the soil, they provided locals with an example of how to successfully cultivate grains on a small scale without slave labor. Quakers' advanced methods of cultivation and application of soil fertilizers such as guano, plaster, clover, and specially prepared manure, even caused grain production to increase during the last decade of the antebellum period. Many ridiculed Virginians' ignorance of advanced farming methods. One northerner insisted that "nothing so provokes a Yankee as the odd way of doing things on a Virginia farm." Despite such condescension, however, locals initially reacted quite positively to the arrival of the Quakers, as they offered new hope for the future of local agriculture. As early as 1847, the *Alexandria Gazette* was pointing out the "beneficial effects" of Yankee labor and capital in Fairfax County, among which were the founding of agricultural societies and an increasing enthusiasm for soil improvement and "scientific farming," which quickly became the new rage.[14]

The Quakers brought with them not only northern farming techniques and methods of soil improvement, however, but also northern ideas about labor. The newcomers were almost unanimously appalled by what they perceived to be the inefficient and backward employment of slave labor in grain production, which they partly blamed for the financial hardships of their slaveholding neighbors. One anonymous visitor to a Quaker farm in western Fairfax observed that in contrast to the local slaveholdings, "here a few free-laborers, prompted by the hope of reward . . . perform all the work that is required; and by doing it promptly and skilfully, the land is improved and brings forth an abundance of the choicest productions." The decline in agriculture among slaveholders, he opined, was doubtless the "inevitable result of its chiefly being performed by a . . . servile population." During his travels in the United States in the 1840s, Scottish geologist Charles Lyell found that the Quakers in Fairfax County had provided "a practical demonstration" that free labor was more profitable than slavery in small-scale grain production.[15]

This sentiment was certainly not shared by many locals, but despite their extreme aversion to the abolitionism of their new neighbors—a source of tension in the region—slaveholders in Fairfax County had long come to accept the hard truth that they simply could not afford to keep surplus slaves. As is clear from table 1.2, between 1810 and 1860 the slave population plummeted from sale, emigration, and manumission (from 5,927 to 3,116, and some 25 percent of the slaves who remained in 1860 were annually hired out to farms and businesses in nearby urban centers). By the eve of the Civil War only 16 percent of the local slave population lived on holdings with more than twenty slaves. While most slaveholders admitted the need for a reduction in their number of slaves, however, few sold *all* of their slaves. The threat of a significant loss in status was no doubt an important factor in this regard. More significantly, not all Virginians were convinced that human bondage was completely incompatible with the new and improved system of agriculture that was taking root. Edmund Ruffin, the Virginia agriculturalist at the forefront of the movement to improve the state's soils, was of the opinion that slave labor could be as capable of producing improved agriculture as free labor, as long as farmers reduced their number of slaves and drove them with greater efficiency. In practice, many farmers strove to keep their number of slaves to a minimum and hire extra hands during especially labor-intensive seasons, such as the harvest, thereby reducing the financial burdens of keeping surplus slaves without giving up their status as slaveholders.[16]

Table 1.2. Population of Fairfax County, Virginia, 1800–1860

Year	Total	Whites	Free Blacks	Slaves	% Slaves in Total Population
1810	13,096	6,626	543	5,927	45 %
1820	11,501	6,224	507	4,770	41 %
1830	9,195	4,892	311	3,992	43 %
1840	9,357	5,469	448	3,440	37 %
1850	10,610	6,835	597	3,178	30 %
1860	11,834	8,046	672	3,116	26 %

Source: U.S. Census, 1810–1860, National Archives and Records Administration.

The transformation of local agriculture which was sparked by the arrival of the Quakers did not radically alter the landscape of Fairfax County overnight, however. The fertile land and successful cotton culture of the Deep South continued to induce small numbers of local farmers to rid themselves of their holdings and make the trek westward and to the south in the 1840s and 1850s, some with their slaves in tow. Emigration had certainly passed its peak by then, but as late as 1846 one slaveholder recorded in his diary that his "neighbor Barlow Mason left his father's residence this day for Louisiana, together with his blacks, to settle there permanently." In 1853, Richard Ashby placed a sale ad in the *Alexandria Gazette* that read: "Having made up my mind to go South, I will sell my farm of 400 acres ... in sight of the Potomac River." Depopulation gave the landscape a depressed and deserted feel. One visitor described the countryside of Fairfax County in the late 1850s as "divided off into the most gloomy looking fields I ever saw, with fences rotting away, houses and out-houses in a state of decay; in short, whole premises on the rapid march of retrogression." By 1860 slave labor had become increasingly marginalized in local agriculture, and small family farms averaging only ninety-four improved acres and a meager cash value of $4,276 prevailed. The county found itself devolving into a society with slaves.[17]

The Empire Built on Rice: Georgetown District, South Carolina

Decades after emancipation, rice planter and master of Woodbourne plantation J. Motte Alston recorded his memories of the slave society into which he was born, describing a region of the rural South in which slaves "out numbered whites nearly one hundred to ten" and the planters "were all fairly rich men." He was hardly exaggerating. The nature and development of regional agriculture in Georgetown District in the South Carolina lowcountry created a world that bore a marked contrast to Fairfax County. Whereas nineteenth-century Fairfax County slipped from its pedestal and began to devolve from a slave society into a society with slaves, Georgetown District, which had long been one of the wealthiest and most firmly rooted slave societies of the South, showed no signs of decline in the antebellum period. The lowcountry not only remained one of the most successful agricultural regions in the slave South, but its highly profitable plantations were so dependent on slave labor that their owners provided some of the leaders of the secession movement that plunged the nation into civil war in 1861.[18]

Rice cultivation in the South Carolina lowcountry was certainly not inevitable but may have been considered as early as the 1670s. High demand for the grain in southern Europe at that time provided a lucrative potential market and early European settlers openly agreed that the lowcountry was particularly well suited for rice, as the physical environment met a number of important requirements for the crop's successful cultivation. First, climatic conditions were ideal. The stifling lowcountry provided a long growing season, averaging between 240 and 300 days a year. Second, the colony had a number of fertile swamps and low-lying tidal lands that provided excellent soil for growing rice. And third, the country offered abundant water sources by which rice fields could easily be irrigated and flooded—a crucial factor in the crop's cultivation. The lowcountry's swamplands and heavy rainfalls during the summer, along with the tide that swelled its coastal rivers and inundated its fertile banks, made it a promising region in which to introduce rice culture.[19]

Despite ideal geographic conditions, however, more than half a century passed before rice as a cash crop firmly took root in South Carolina. Early European settlers lacked a thorough knowledge of rice cultivation and failed to master its techniques for several decades—"the mysteries of cultivation were not unravelled quickly," in the words of historian Peter Wood. Initial experiments with the grain indeed met with little success. When in the late 1680s and early 1690s a better strain of seed was introduced from Madagascar, colonists redoubled their efforts. This time, after much trial and error, they finally began to see results—although even after they had successfully mastered the crop's cultivation by around 1700, they remained ignorant of its cleaning and processing for years. As late as 1720, when the Chesapeake was already a mature tobacco-based slave society, the South Carolina economy depended more on mixed farming, cattle raising, and naval stores than it did on rice. In the decade that followed, however, rice took off as a major cash crop in the lowcountry. Exports tripled in the 1720s, and in the 1730s expansion reached a hectic pace as more land was cleared and planted in rice than in the previous forty years combined. Dubbed "Carolina gold" by those who stood to profit from it most, the new staple—supplemented by indigo—quickly transformed the landscape, economy, and demography of the lowcountry.[20]

Closely paralleling the rise of rice as a staple crop in South Carolina was the exponential growth of its slave population, as planters imported African slaves by the thousands to clear the land, dig canals and irrigation ditches, build earthen banks, and toil in the muddy rice fields. Although South Carolina was a slave society from the outset—the first English settlers from Barbados had brought their slaves with them—the lack of a specific staple crop limited colonists from importing especially large numbers of bondspeople in the late seventeenth century. When large-scale rice cultivation was finally mastered, however, aspiring planters stormed Charleston's slave auctions, sometimes buying groups of thirty or forty slaves at a time. In the 1730s alone, rice planters imported approximately 21,150 slaves. Between 1700 and 1780, more than 110,000 victims of the Atlantic slave trade landed at Charleston and were subsequently auctioned off to live out the rest of their lives in bondage on nearby rice plantations. Before long, blacks formed a majority of the population in the colony, a demographic composition unique to South Carolina. Indeed, visitors to the region frequently commented that "Carolina looks more like a negro country than like a country settled by white people."[21]

Many factors contributed to the rise of slave labor in eighteenth-century South Carolina, but two in particular seem to have proved crucial to the success of rice in the lowcountry. First, many enslaved people brought to the lowcountry from West Africa were broadly familiar with rice planting, in contrast to European settlers. The majority of lowcountry slaves arrived from Angola, the Windward Coast, the Gold Coast, and Gambia—all regions in which rice was to some extent cultivated locally and often sold to European traders. Certainly not all—not even a majority—of the newly arrived African slaves had a technical knowledge of rice planting. Especially those originally from the interior had probably never seen a rice plant, and Philip Morgan has argued moreover that in most West African cultures rice cultivation was primarily the responsibility of women, not men. Nevertheless, in a region where planters were ignorant of many aspects of rice culture, acquiring a labor force of which at least some members were familiar with the crop's cultivation was an obvious advantage. Indeed, numerous scholars have pointed out that enslaved people who were already familiar with rice cultivation in their homelands retained a number of African traditions in several features of the rice cycle until well into the nineteenth century.[22]

Enslaved people from West Africa also appeared to be largely resistant to certain diseases that repeatedly ravaged the unhealthy rice swamps of the lowcountry. Yellow fever, a disease to which most European settlers had never been exposed, recurred regularly in South Carolina, taking a heavy toll on the white population. The disease was widespread along much of the West African coast, however, and so enslaved people coming from that region were more likely to have acquired immunity. Likewise, West Africans were less likely to suffer from chronic malaria, a disease that also plagued the Carolinas during the summer months. These factors may not have been the primary reasons for employing African slave labor in rice cultivation, but they certainly reinforced rice planters' decision to do so and contributed significantly to the slave-buying frenzy that led to a black majority in the eighteenth century.[23]

One of the chief reasons for the great expansion of rice in the 1730s was the opening up of virgin lands for settlement in Georgetown District, north of Charleston. A coastal district traversed by five Winyah Bay tributaries—the Sampit, Black, Pee Dee, Waccamaw, and Santee rivers—Georgetown was ideally situated for the establishment of large rice plantations, as frontage on these waterways proved especially advantageous to successful rice planting. Whereas an earlier generation of planters experimented upriver with dams and water reserves to obtain the water necessary to flood their rice fields, both with varying degrees of success, a new generation of planters along the broad tidal estuaries of Georgetown District discovered that the natural rise and fall of the tide allowed them to control flooding and drainage relatively easily, an advantage that increased productivity and shortened the growing season. Moreover, pioneering Georgetown planters such as Thomas Lynch and John and William Allston were able to profit from the past experiments of their lowcountry neighbors and could tap into the ever-growing slave labor supply that was pouring into Charleston. They met with an uncanny success, and the rice and indigo plantations—the latter crop initially adding to the large fortunes being made in the Winyah Bay area, even leading to the establishment of the aristocratic Winyah Indigo Society—expanded rapidly as a result. As early as the mid-eighteenth century, men and women in bondage afforded local rice planters the highest per capita income in the American colonies. By the outbreak of the American Revolution, more than half of the enslaved population of Georgetown District lived on plantations with over fifty slaves, and over one-fifth lived on holdings with one hundred slaves or more, dwarfing most slaveholdings in northern Virginia.[24]

Time would tell that lowcountry planters had invested wisely, as the rice industry proved remarkably stable in the long run. Unlike tobacco in Fairfax County, rice survived the Revolutionary War, although the hostilities of course took their toll on local agriculture. It certainly helped that rice planters had enjoyed a long period of prosperity before the war broke out. Rice did not exhaust the soil, as tobacco did, and the level landscape prevented significant erosion. Nevertheless, the war temporarily halted the prosperity of local planters, especially when the Continental Congress imposed an embargo on exports of provisions, including rice, in 1777. Cut off from most of their European markets, lowcountry planters struggled during the first couple of years to find markets closer to home for their crop, some of it finding its way to the French West Indies (with the approval of Congress) and some of it to the Continental Army. Produce from Plowden Weston's plantation in Georgetown District, for example, supplied the troops of Peter Horry during the war. For most planters, however, an unlikely stroke of luck arrived when the British invaded South Carolina in 1779. The British occupation of Charleston and the lowcountry during the last four years of the war allowed rice planters to ignore the embargo and afforded them with an outlet for their staple crop. Disorganization and property damage prevented normal production, but trade was sufficient for one contemporary to describe Charleston harbor in 1781 as "pretty well filled with shipping which affords a comfortable appearance, of goods imported & Rice & Indico to be exported."[25]

What disrupted the economy of the South Carolina lowcountry most during the war was a significant loss in its enslaved population. The confusion and loss of authority that accompanied the war in the region provided enslaved people with many opportunities to escape bondage on the rice plantations. The promise of freedom lured many to seek asylum with, and fight for, the British. Indeed, when the British evacuated Savannah and Charleston at the end of the war, some ten thousand blacks left with them. Other groups of runaway slaves formed maroon societies in the swamplands. Scholars have estimated that South Carolina as a whole lost as much as 30 percent of its enslaved population to flight, migration, and death during the Revolution, a significant decline by any standard. The loss of labor suffered by rice planters in Georgetown District specifically is unknown, but after peace was restored, evidence indicates that Georgetown planters made up for whatever losses they sustained by importing more slaves from both foreign and domestic traders, just as they had in the years prior to the war.[26]

The rice industry in Georgetown District recovered quickly from the effects of the Revolutionary War, but indigo collapsed and was virtually eliminated from the lowcountry by the turn of the nineteenth century, a result of falling prices and ravaging insects. At a time when Fairfax County planters were losing their primary staple and intensifying their production of secondary staples, thus, planters in Georgetown District were experiencing the opposite—the loss of their secondary staple encouraged them to concentrate more than ever on rice. By 1800, planters in Georgetown District had already significantly expanded and consolidated their operations, as the geographic center of the rice industry shifted from the inland swamps to the more efficient river plantations near the coast. The result of this shift was a dramatic growth in plantation size and the virtual elimination of inland competition; smaller planters increasingly sold out to the larger planters along the rivers and emigrated. In the opening decade of the nineteenth century alone the proportion of local slaves living on holdings with more than one hundred slaves increased by 14 percent, from 44 percent in 1800 to 58 percent in 1810. The proportion of slaves living on holdings with more than two hundred slaves likewise jumped by 14 percent, from 16 percent in 1800 to 30 percent in 1810. One nineteenth-century resident claimed that her neighbors were interested only in "mak[ing] Rice to buy Negroes and Buy[ing] Negroes to make Rice."[27]

As Georgetown's grandees extended their holdings, the American rice industry became consolidated in their hands. Production in the district during the first half of the nineteenth century seemed limitless and increased steadily throughout the antebellum period, at one point even accounting for 45 percent of the national total. Indeed, despite increased competition due to the geographic expansion of the rice industry into the Gulf States during the last two decades of the antebellum period, rice production in Georgetown District increased each decade by roughly ten million pounds (see table 1.3). By the time the Civil War broke out the district resembled, in the words of one local historian, a virtual "kingdom built on rice—an empire that sprawled over thousands and thousands of acres."[28]

Table 1.3. Rice Production in Georgetown District, South Carolina, and the United States, 1840–1860

	1840		1850		1860	
	Pounds	% of Total	Pounds	% of Total	Pounds	% of Total
Georgetown District	36,360,000	45	46,765,040	22	55,805,385	30
South Carolina	60,590,861	75	159,930,613	74	119,100,528	64
United States	80,841,422	—	215,313,497	—	187,167,032	—

Source: Compiled from Rogers, *The History of Georgetown County*, 324; Gray, *History of Agriculture*, 2:723.

That "empire" was—more than any other region of the South—predominantly a black world. A majority of slaves in Georgetown lived on holdings with more than one hundred slaves, and enslaved people consistently formed between 85 percent and 89 percent of the total population between 1810 and 1860 (see table 1.4). Living on such vast plantations in a region where they outnumbered whites almost nine to one, most enslaved people rarely came into contact with any whites other than their overseers. William Howard Russell, a British reporter for the *Times* who visited the district when the Civil War broke out, was surprised to see that the slave children were so unused to seeing whites that "they generally fled at our approach. The men and women were apathetic, neither seeking nor shunning us, and I found that their master knew nothing about them." Elizabeth W. Allston Pringle, who was the daughter of South Carolina governor Robert F. W. Allston and grew up on Chicora Wood plantation on the Pee Dee, remembered long after emancipation that "in the low-country, where each plantation had a hundred or more negroes, which necessitated separate villages . . . the negroes lived more or less to themselves." Such was the distance between the big house and the slave villages that when her parents visited their bondspeople at all they would have to get "into the buggy and [drive] off down the beautiful avenue of live oaks, draped with gray moss, out to the negro quarters, which is always called by them 'the street.'"[29]

Exploiting the labor of their massive and largely anonymous slave populations, the tiny minority of planters in Georgetown catapulted themselves into the apex of the South's wealthy slaveholding elite, their plantations producing vast quantities of their profitable staple crop. Keithfield, a plantation owned by Richard O. Anderson, had a workforce of 384 slaves and

Table 1.4. Population of Georgetown District, South Carolina, 1800–1860

Year	Total	Whites	Free Blacks	Slaves	% Slaves in Total Population
1800	15,962	4,055	91	11,816	74 %
1810	15,602	1,710	102	13,790	88 %
1820	17,604	1,830	227	15,547	88 %
1830	19,872	1,931	214	17,727	89 %
1840	18,299	2,093	188	16,018	88 %
1850	20,288	2,193	201	17,894	88 %
1860	21,230	3,013	183	18,034	85 %

Source: U.S. Census, 1800–1860, National Archives and Records Administration.

produced more than 1.1 million pounds of rice in 1850. Martha Allston Pyatt, the widow of rice planter John Francis Pyatt, produced over two million pounds of rice in 1850 with a staggering labor force of 768 slaves, who were spread out over four plantations. Few could compare with the grandeur of Joshua John Ward's estate. Dubbed "the king of the rice planters," Ward was the largest rice producer and slaveholder in the United States. His 1,092 slaves produced over 4.4 million pounds of rice in 1850, grossing over $141,120 for their master in one year. One of Ward's former slaves, Ben Horry, was hardly exaggerating when he exclaimed to interviewers of the Federal Writers' Project: "Rice been money in dem days!"[30]

Georgetown District and the rice industry were inseparable in the minds of contemporaries throughout the antebellum period, as the crop dominated virtually every aspect of life along the Winyah Bay tributaries. A traveler to the district in the 1820s reported that "every thing is fed on rice; horses and cattle eat the straw and bran; hogs, fowls, etc. are sustained by the refuse; and man subsists on the marrow of the grain." For planters, subsistence came more from the huge profits than from the marrow of the grain. Underpinned by a highly successful agricultural sector, Georgetown District remained the very definition of a slave society right up until emancipation.[31]

A Forced Crop: St. James Parish, Louisiana

If Fairfax County represents a devolving slave society in the nineteenth century, and Georgetown District a stable and even expanding one, then St. James Parish in southern Louisiana provides scholars with an excellent example of a rapidly developing slave society during the era of feverish expansion in the antebellum Deep South. Slavery may have been a marginal institution to southern Louisiana's original French and Spanish settlers, but the introduction and expansion of sugar cultivation in the late 1790s, as well as the territory's subsequent acquisition by the United States in 1803 and admission as a state in 1812, triggered a rapid transformation into an American-style slave society, as the region experienced an economic and demographic boom that was halted only by the outbreak of the Civil War. At the dawn of the nineteenth century, St. James, a small parish situated halfway between Baton Rouge and New Orleans and straddling the Mississippi River, found itself at the epicenter of the American sugar industry.[32]

Sugarcane was not only an unlikely candidate for cultivation in St. James Parish, but it was an unlikely candidate for Louisiana or anywhere else on the North American mainland. Indeed, ideal conditions for the cultivation of the crop—warm temperatures year-round with no frost, a well-distributed annual rainfall, an average growing season of eighteen to twenty months—were to be found not along the Gulf Coast but rather in the West Indies and Brazil. Sugarcane especially flourished on the islands of the Caribbean, becoming the major slave-based staple crop of the Americas. The absence of winter freezes there allowed sugar to reach full maturity and harvests to be undertaken without hurry. Moreover, cane growth from old roots—called "ratoon growth"—often continued in the tropical Caribbean for up to twelve years in a row.[33]

In semitropical southern Louisiana, and especially in the Mississippi River parishes such as St. James, however, ideal natural conditions for the cultivation of sugarcane were absent. There existed a "lack of complete harmony between land and product," in the words of historian J. Carlyle Sitterson. Icy winds and winter frosts in Louisiana annually threatened to destroy standing cane, causing the juice to deteriorate and thus shortening the growing season to an average of only nine to ten months instead of eighteen to twenty. The colder climate forced sugar planters to harvest their crop at a frenzied pace lest it be ruined by a sudden autumn frost, and ratoon growth could be expected for only three years at most, requiring more frequent replanting than in the West Indies. Annual rainfall in the region, while plentiful, is hardly well distributed throughout the year. Indeed, in 1844 two severe droughts very nearly destroyed an entire sugar crop in St. James Parish. "The cultivation of sugar cane in Louisiana is only a forced one," reported J. A. Leon, a nineteenth-century researcher of the sugar industry in Louisiana and the British colonies.[34]

Despite the climatic handicaps for cultivating cane, numerous incentives for commercial agriculture did exist in southern Louisiana that inspired early French and Spanish settlers in the eighteenth century to experiment with a number of other potential cash crops. The soil along the Mississippi River, which bisects St. James Parish, was exceptionally fertile—"it may be said inexhaustible," according to one observer—and frontage on the Mississippi itself provided would-be planters with a natural means of transporting their crops to New Orleans and abroad. Much like South Carolina in the early decades of the eighteenth century, however, southern Louisiana failed to adopt a single staple crop for decades. As late as the 1780s a variety of commodities were being produced in the region, including "indigo, tobacco, timber, cotton, pitch tar, rice, maize, and all kinds of vegetables," but none had become a dominant cash crop, and few were cultivated on a scale that would necessitate much slave labor. The only crop that came close was indigo, cultivated by some on a relatively large scale until the end of the eighteenth century. (The first plantation in St. James Parish was an indigo plantation called Cabahannocer, owned by the Cantrelle family.) However, Louisiana indigo failed to compete with indigo from South Carolina.[35]

Experimentation with sugarcane as a potential cash crop began earlier than one might expect, considering the disadvantages of the Louisiana climate. In the 1750s and 1760s a handful of planters living near New Orleans attempted to cultivate cane, but they failed to establish the sugar industry in Louisiana on a commercial basis. Like their counterparts in the lowcountry, early planters in southern Louisiana failed to unravel the mysteries of cultivation of what would later become their staple crop. The shortened growing season resulted in an inferior product that often failed to properly granulate, a predicament exacerbated by planters' lack of adequate skills and machinery to produce sugar. Indeed, determining when the syrup was ready to granulate proved so tricky that early planters became convinced that they would always have to sell only molasses. One small shipment of Louisiana sugar sent to France in 1765 was so imperfectly granulated that it leaked out of the containers en route. Moreover, few planters believed that they would ever really be able to compete with the British, French, and Spanish sugar islands, which amply supplied the world's demand for sugar.[36]

Developments in the 1790s, however, changed everything. First, the indigo industry began to collapse as a result of falling prices and ruinous insects, which destroyed entire crops and wiped out potential profits at virtually the same time that they did in the South Carolina lowcountry. Second, the slave insurrection of Saint-Domingue in 1791 reduced North America's supply of sugar, raised its price, and sent hundreds of skilled sugar makers fleeing to Louisiana. Faced with financial ruin and enticed by the prospects of filling a new gap in the market, a handful of Louisiana planters immediately renewed their experiments with cane, initiating a major economic shift just as planters in Fairfax were shifting to grains and planters in Georgetown were tightening their hold on the rice industry along the Eastern seaboard.[37]

Only when Etienne de Boré finally succeeded in granulating sugar in 1795, however, did the southern Louisiana sugar industry truly take off in earnest. In 1794 De Boré had borrowed an initial cash outlay of $4,000; constructed a sugar mill, drying room, and shed; and employed the labor of forty slaves to plant his fields with Creole seed cane, the only variety available at the time. Meticulously following his crop's progress, he had carefully adapted his methods of cultivation to avoid the dangers of the Louisiana climate. During droughts in the spring he irrigated his fields to speed up the maturation of the cane before the killing frosts of autumn arrived. When autumn came, he decided against harvesting his crop all at once and storing it to be milled during the winter months—a practice that often resulted in the souring of the cane. Instead, he milled it as soon as it was cut to ensure maximum quality, leaving the rest of his crop densely packed in the fields to prevent it from frost damage until it too could be cut and milled. To supervise the risky business of boiling and granulating sugar, De Boré had hired an experienced sugar maker from Saint-Domingue named Antoine Morin, who effectively announced the commencement of southern Louisiana's golden age when he famously cried out, "It granulates!" amidst a crowd of onlookers. De Boré sold his crop in 1796 for $12,000, clearing over $5,000 in profit and paving the way for the Louisiana sugar boom of the nineteenth century.[38]

Immediately following De Boré's success, planters up and down the Mississippi River shifted to commercial cane cultivation—no small task considering the expenses involved in setting up a sugar plantation. Sugar plantations not only had to grow cane but manufacture granulated sugar and molasses as well. Thousands of dollars were required to construct mills, purchase boilers, hire experienced sugar makers, and purchase slaves. One contemporary reported that "To begin a new plantation, money must be procured to buy negroes, land, provisions, materials for extensive buildings, machinery, and for the daily expenses of the mechanics; clearing the land, its tilling, planting, making the many hundred thousands of bricks for the sugar house, curing house, hospital, negroes' houses, dwelling house, &c., &c., and two years of very hard work are spent before producing the first pound of sugar." Setup costs only increased with time. Frederick Law Olmsted, a reporter for the *New York Daily Times* who journeyed through the South between 1852 and 1857, learned during his visit to southern Louisiana that "the capital invested in a sugar plantation . . . ought not to be less than $150,000."[39]

Then there were the risks to consider. As Olmsted correctly cautioned, "if three or four bad crops follow one another, [the sugar planter] is ruined." Planters' anxieties concerning natural disasters were certainly justified. Frost damage has already been mentioned as a cause of considerable worry for Louisiana sugar planters, but there were other dangers as well. Destructive hurricanes severely injured the sugar crops of 1812, 1824, 1832,

1856, and 1860. And the Mississippi itself was prone to frequent flooding and sometimes broke through the levees, often wiping considerable acreage of alluvial land off the map. J. A. Leon commented during his residence in Louisiana that "the overflow of the Mississippi river frequently breaks the banks and inundates the fields; whole crops are lost by those disasters." Traveler Victor Tixier reported such a disaster in St. James when "in one single night the river tore away two *arpents* of land and came very near the house." (An arpent was the French measure of land, equal to 0.85 of an acre.) Despite the high costs, substantial risks and hard work, however, the number of sugar plantations in southern Louisiana mushroomed from ten in 1796 to more than 1,300 by the outbreak of the Civil War. As one historian put it, "the prize was worth the risk," and the sugar planters were simply not to be discouraged.[40]

The greatest obstacle for planters at the turn of the nineteenth century was acquiring a slave labor force, which proved especially costly in the first few years as African and Creole slaves (slaves born in Louisiana) were severely limited in number.[41] Before the sugar boom southern Louisiana had been a classic society with slaves, with slave labor constituting only a marginal institution within the colony's economy. Despite repeated attempts by the French and Spanish Crowns to create a slave colony in the Lower Mississippi Valley, human bondage never became especially widespread due to the lack of a lucrative cash crop. Indeed, slavery was a predominantly urban institution under both French and Spanish rule. Many enslaved people in eighteenth-century New Orleans were eventually manumitted or allowed to hire themselves out, and to contemporaries human bondage seemed like a system that was unlikely to take root as it had in the Caribbean or elsewhere in North America. The destructive slave revolt on Saint-Domingue and subsequent local conspiracy by slaves at Pointe Coupée to rise up against their masters convinced Spanish colonial authorities to even ban the importation of African slaves in the 1790s.[42]

De Boré's successful experiment with sugar transformed the demand for slave labor, however, which skyrocketed almost overnight. And with the recent ban on slave imports driving prices for local field hands as high as $1,200, it was clear that something had to be done to supply the colony with a suitable labor force. The solution came when the United States purchased the Louisiana Territory in 1803. An act passed by Congress in 1805, which banned the importation of slaves into Louisiana if they had been brought from Africa after 1798, was conveniently interpreted by local authorities to permit the importation of slaves from any other part of the United States (including those born in Africa), opening the floodgates to a flourishing slave trade that continued on a large scale until the Civil War. For new sugar planters, setup costs were already so high that many were forced to make do with a limited number of slaves until their plantations began to pay for themselves. Most found financial success quickly forthcoming, however, sustaining an almost constant demand for slaves as plantations expanded their operations. The slave population in the sugar parishes grew from under 10,000 in 1810 to over 88,000 in 1860, largely from importation. Advertisements for large shipments such as the following were a common sight in New Orleans newspapers: "NEGROES FOR SALE. About one hundred Virginia slaves, of both sexes, at the corner of Dauphin and Canal Streets," and "Great Excitement!!! 400 slaves expected to arrive by the first of November . . . comprised of every size, age, and sex to suit the most critical observer."[43]

With the supply of slave labor secured, a number of other developments during the first half of the nineteenth century served to further stimulate the growth of the sugar industry. First, the U.S. acquisition of Louisiana provided sugar planters with a lucrative domestic market and attractive government tariff protection from West Indian competition. This was of major importance because Louisiana sugar was grown almost exclusively for domestic consumption. Second, experiments with different varieties of seed cane in the early nineteenth century proved fruitful, resulting in the adoption of "ribbon" cane—a variety that matured rapidly and was more resistant to frost than the varieties used by De Boré and his successors. Finally, productivity was increased with a number of technological innovations. In 1822 the first horse- and oxen-powered sugar mills were replaced by steam-powered mills, a considerable improvement that was quickly adopted throughout the region. The introduction of vacuum pans in 1830 by Gordon and Forstall, and Valcour Aime (planters from St. James Parish) added to the giant leap in technology. And by the eve of the Civil War, sugar planters boasted the most advanced evaporators and machinery, and a few of the wealthiest even had refineries.[44]

All of these factors caused sugar production and productivity to increase exponentially throughout the nineteenth century, putting southern Louisiana on the map in the slave South. As early as 1829, planters were reportedly producing an average of 3,000 to 4,000 pounds of sugar per hand. Annual output soared from 30,000 hogsheads in 1823 to over 460,000 hogsheads in 1861 (one hogshead equaled roughly 1,000 lbs). Enticed by tariff protection, technological advances, and especially the substantial profits being made by planters along the Mississippi River—"the facility with which the sugar Planters amass wealth is almost incredible," were the sentiments of one visitor—thousands of Anglo-Americans from the struggling Upper South came pouring into southern Louisiana with high hopes of making their fortunes. One traveler reported that "the Americans are coming down in droves from the north of the United States . . . they flock over all of Louisiana in the same way as the holy tribes once swarmed over the land of Canaan." As late as the Civil War, newcomers came to establish themselves in the Louisiana sugar country, by then a region engulfing not only the Mississippi River parishes south of Baton Rouge but bayous Lafourche and Teche as well.[45]

In St. James Parish, ideally located in the heart of the sugar country between fifty and seventy-five miles north of New Orleans by river, sugar had taken over as early as 1803, and developments closely paralleled those of the Louisiana sugar region as a whole. The first census of the parish, taken in 1766, listed only sixteen slaves, indicating a society with slaves for which bondage was for all intents and purposes a nonexistent institution. By 1810 there were 1,755 slaves, however, and by 1860 the slave population had increased to some 8,090 men and women, with slaves outnumbering whites by a factor of two to one (see table 1.5) and comprising some $4 million of invested capital in local agriculture. The 123 slaves living on W. P. Welham's Homestead plantation in 1860 alone were valued at over $100,000. One traveler visited a plantation in St. James Parish in 1852 whose host had more than $170,000 invested in 215 slaves. John Burnside, exceeded only by Georgetown rice planter Joshua John Ward in wealth and slave ownership in the South, owned 186 slaves on his plantation in St. James and another 754 on holdings in neighboring Ascension Parish in 1860, representing a personal property value of $2.6 million. The sugar masters in St. James turned narrow estates into agricultural empires—averaging 778 improved and 1,112 unimproved acres by the eve of the Civil War—and reinvested their large profits into clearing more land and buying more slaves.[46]

Annual output in the parish likewise reached unprecedented levels by the end of the antebellum period. Valcour Aime produced a respectable 112 hogsheads of sugar when he began operations on his plantation in 1823. Three decades later, however, during the banner year of 1853, he had increased his annual production to 1,064 hogsheads, netting over $73,000. Orange Grove plantation produced 900 hogsheads in 1853 "with only 106 negroes, old and young, men, women, and children included . . . [and] five of [whom] walk on wooden legs." William Webb Wilkins, also from St. James, produced 900 hogsheads that year with only sixty able-bodied field hands. While sugar production varied from year to year and setbacks were frequent due to unexpected frosts, hurricanes, and floods—which caused damaged crops in 1850–51, 1855–56, 1856–57, and 1859–60, for example—planters in St. James Parish were generally producing large sugar crops by the end of the antebellum period (see table 1.6).[47]

The wealth and grandeur of established local sugar plantations were frequently noted by visitors and passers-by alike. T. B. Thorpe, who traveled to St. James Parish in 1853, remarked that "the stranger who for the first time courses the 'Father of Waters' at a season of the year . . . [and] looks over and down upon the rich sugar plantations, is filled with amazement, and gets an idea of agricultural wealth and profuseness nowhere else to be witnessed in the world." Amos Parker, another traveler, reported seeing "armies of negroes" at work in the vast sugar fields of St. James Parish in the

Table 1.5. Population of St. James Parish, Louisiana, 1810–1860

Year	Total	Whites	Free Blacks	Slaves	% Slaves in Total Population
1810	3,692	1,896	41	1,755	48 %
1820	5,655	2,522	52	3,081	55 %
1830	7,607	2,557	26	5,024	66 %
1840	8,541	2,762	75	5,704	67 %
1850	11,071	3,285	62	7,724	70 %
1860	11,537	3,348	61	8,128	70 %

Source: U.S. Census, 1810–1860, National Archives and Records Administration.

Table 1.6. Annual Sugar Production (in hogsheads) in St. James Parish and Louisiana, 1849–1860

Year	St. James Parish	Louisiana	% St. James Crop of State Total
1849–50	21,978	247,923	9 %
1850–51	14,936	211,201	7 %
1851–52	17,719	237,547	7 %
1852–53	24,417	321,934	8 %
1853–54	33,736	449,324	8 %
1854–55	25,560	346,635	7 %
1855–56	16,331	231,429	7 %
1856–57	7,665	73,296	10 %
1857–58	19,022	279,697	7 %
1858–59	26,749	362,296	7 %
1859–60	15,400	221,840	7 %

Source: Compiled from Champomier, *Statement of the Sugar Crop in Louisiana*, 1849–60; Gray, *History of Agriculture*, 2:1033.

1830s, while their owners relaxed in "elegant houses, surrounded by orange trees, loaded with fruit." Victor Tixier observed during a visit to St. James that the sugar planters "lead a pleasant life indeed, served according to their wishes, living in abundance and even luxury." Visitor Philo Tower admitted that "here my ideal of southern grandeur, wealth, magnificence, &c., were not only fully realized but a little exceeded." And J. W. Dorr was deeply impressed by the wealth of St. James Parish sugar planters in 1860: "In this parish are many very extensive sugar estates . . . ranging from [$400,000] up among the millions. The residences on most of these plantations are really magnificent . . . and the grounds are a paradise of rare and beautiful plants, trees and shrubs."[48]

Conclusion

Although these three agricultural communities all constituted slave societies during the nineteenth century, they each found themselves in different stages of development. The transition from tobacco to mixed grains in Fairfax County did not adequately solve planters' financial problems, which gradually worsened. Enslaved people in northern Virginia thus found themselves under the precarious ownership of frustrated and increasingly indebted masters for whom slave labor was a financial burden more than anything else; the region slowly began to devolve in the direction of a society with slaves. The rice plantations in Georgetown District were far more stable. With a thriving market for "Carolina gold," rice planters were able to increase their land and slaveholdings as the nineteenth century progressed. Antebellum slaves in Georgetown District lived on large, stable, monoculture plantations during an era of incredible economic success. By any definition, Georgetown was and remained a strong slave society. As sugar took root in southern Louisiana at the turn of the nineteenth century, St. James experienced an unprecedented economic and demographic boom. Profits soared, plantations expanded, and the slave population increased exponentially, turning the once quiet river parish into a thriving American slave society. Enslaved people in St James Parish lived on ever-expanding and increasingly wealthy plantations throughout the nineteenth century.

As will become clear in parts II and III, the success or failure of slave-based agriculture in each of these three regions significantly affected the boundaries and opportunities with which slave families in the nineteenth century were confronted, resulting in drastically different experiences for enslaved people across time and space.

II

The Balancing Act

Work and Families

2

The Nature of Agricultural Labor

The nature of agricultural labor in various southern localities had important consequences for enslaved people's time and flexibility in reconciling their status as forced laborers with their duties as family members. Few scholars would disagree that work defined time in the rural slave societies of the nineteenth-century South and was thus the most important organizational factor in the experiences of enslaved people. At the daily and seasonal levels, methods of cash crop cultivation and prescribed work patterns in the fields determined the extent of enslaved people's contact with family members, both during working hours and in private. The nature of work moreover determined the boundaries and opportunities with which slaves were confronted with regard to child rearing, domestic responsibilities, and family-based internal production.

This chapter explores and explains the daily and seasonal work of enslaved people in each of the three chosen regions of the non-cotton South. The aim of this chapter is to provide a broad understanding of enslaved people's agricultural labor, as well as a basis from which to further examine the boundaries and opportunities created by work for slave families in the following two chapters.

Jacks-of-All-Trades: Fairfax County, Virginia

Frank Bell, who spent his youth in bondage on a wheat plantation near Vienna, belonged to the last generation of the Bell family forced to toil in the fields of northern Virginia as slaves. Bell was twenty-six years old when the Civil War broke out, but, as he later told interviewers, his "pappy and my grandpappy wukked for ole Marser's people all dey lives." His grandfather Starling died an old man some fifteen years before the war (Bell claimed that he lived to be over a hundred years old), and therefore he must have witnessed firsthand the fateful transition from tobacco to mixed grains at the turn of the nineteenth century. However, while cultivation methods changed significantly, Starling most likely found more continuity in general work patterns than one might expect. Although they cultivated different cash crops from those grown by their eighteenth-century forebears, the enslaved people on the wheat and corn farms of antebellum Fairfax County employed work patterns that were directly inherited from—and sharply influenced by—the old tobacco culture. Indeed, as tobacco was still cultivated on an ad hoc basis by a handful of planters during the first half of the nineteenth century, some slaves found themselves engaged in the cultivation of mixed grains *and* tobacco. Both crops placed substantial demands on slave laborers.[1]

For eighteenth- and early nineteenth-century field hands employed in the cultivation of tobacco, the planting cycle began in January, when land was cleared or burned and the soil was prepared for early planting.[2] Between the end of February and the beginning of March, slaves began to carefully sow tobacco seeds in specially prepared beds of mulch. This was an extremely fragile stage, as a late frost could kill the young seedlings, requiring field hands to cover seedlings with leaves to protect them from the cold. By the beginning of April, transplanting could begin, and the workload began to steadily increase. New fields were cleared and prepared with thousands of small hills to receive the transplanted tobacco plants. Wielding the hoe, most able-bodied slaves were expected to make at least 350 hills in a day. During the summer months, field hands were kept busy with weeding, transplanting, and replanting. They also rid the fragile tobacco leaves of ravaging caterpillars—a seemingly endless chore.

By the end of August, the hoes were laid by and the first tobacco plants were ready for harvesting. Because not all tobacco plants ripened at the same time, the harvest was often dragged out from August through September and into early October. Once the leaves were harvested, slaves hung the cut tobacco from the rafters in the tobacco house, allowing the leaves to fully dry before stripping them from the stalks and individually rolling and packing them into hogsheads. This work was tedious and required long hours, patience, careful handling, and close supervision, as the tobacco plants remained delicate and fragile. While the harvest was usually finished by October, duties such as drying, stripping, and packing kept field hands busy through December, after which the cycle repeated itself.

Tobacco, according to one eighteenth-century planter, was "a plant of perpetual trouble and difficulty." Its fragile and unpredictable nature, tedious cultivation, and variable quality required a maximum amount of attention and supervision by both planters and slaves. Partly as a result of this, tobacco planters in Fairfax County and elsewhere in the Chesapeake organized the work of their enslaved field hands by the gang—or time-work—system. In other words, enslaved people were set to work for a fixed amount of time, not unlike the work of most enslaved people in the Americas. They labored from first daylight in the morning until sundown in the evening, usually in small, interdependent squads where their progress could be closely monitored by the planter or overseer. George Washington required his hands at Mount Vernon to "be at their work as soon as it is light, [and] work till it is dark." During the winter months, men and women in bondage often even worked by candlelight into the evening hours, curing, stripping, and packing tobacco.[3]

The transition from tobacco to wheat, corn, and other small grains at the turn of the nineteenth century, however, placed very different demands on enslaved laborers. New methods of cultivation were learned, a new work rhythm was implemented, and the tobacco calendar, which defined time in the world of northern Virginia slave families, was altered as a variety of winter and summer crops were introduced with different seasonal cycles, keeping field hands busy throughout most of the year with continually alternating responsibilities.

The farm journal of David Wilson Scott, a Fairfax County slaveholder who owned nineteen bondspeople in 1820, offers insight into the local cultivation of grains. The agricultural calendar for Scott's workforce began not in January, as it did on tobacco plantations, but in late summer and early autumn, when the farm's winter crops—wheat and rye—were planted. In the year 1819, enslaved people on Scott's farm began early. On 31 August they went to the fields to prepare the land for wheat planting by plowing and hoeing for just over a month. On 2 October they began to sow, plow, and harrow in wheat on different lots throughout the estate, a labor-intensive duty that kept them busy until 25 October. The custom was to sow "broadcast," then drag "a heavy harrow over it," according to one local farmer. The last week of October was spent sowing small quantities of rye.[4]

That done, the slaves immediately shifted their attention to the cornfields, the harvest of which consumed Scott's field hands throughout November and into the first week of December. Corn was harvested by "topping": the stalks were cut off just above the ears, and after the corn was ripe and fully dried, the ears were pulled from the stalks and brought to the barn, where they were husked. The field hands also fertilized the recently planted wheat fields by sowing in plaster at this time. (Scott was a progressive planter; many of his neighbors did not adequately fertilize their fields in the 1820s.)[5]

The winter months ushered in a relatively slow period in the agricultural calendar, filled in by a myriad of odd jobs. One of the most pressing duties that enslaved people performed in the winter was treading out the wheat that had been harvested earlier in the year. This was usually done in the barn on a "treading floor" measuring forty feet square. A ring of sheaves four feet wide was laid around the floor, and two horses were set to walking around on it while a couple of hands continually turned it over with pitchforks. John Jay Janney, a farmer from neighboring Loudoun County, claimed that "treading out wheat always made me feel as if I had a cold, headache, back-ache and slight fever, the result of the dust." When the grain was all out of the straw, it was run through a wheat fan to separate the wheat from the chaff. Some farmers in the late antebellum period owned threshing machines. Christopher Nichols, once enslaved in northern Virginia, remembered having to "stand before the drum of the wheat machine, and tend the machine all day" during the winter. Other winter tasks included threshing rye with a flail; grinding corn and wheat, or taking it to the local mill to be ground; chopping wood; repairing fences; building stalls for the livestock; slaughtering hogs; and, in the case of Scott's slaves, smoking and preparing almost two thousand pounds of pork.[6]

By the beginning of February, Scott's field hands began to plant again, this time in the farm's provision grounds and orchards. Potatoes, peas, lettuce, and cabbage were planted, as well as a number of fruit trees, mostly peach and apple. At the same time, other hands were employed plowing and preparing the soil for the summer crops, corn and oats. During the first half of April, some slaves were employed in the fields planting oats, while others were sent back to the provision grounds to plant carrots, parsnips, beets, and pumpkins. From 15 April on, however, almost all hands were employed planting corn, a labor-intensive occupation that required slaves to make thousands of small hills with their hoes to receive the corn seeds. On Saturday, 24 April 1819, Scott proudly recorded in his diary that his hands had "planted about 19,000 Corn hills." Corn planting lasted until 7 May, after which his enslaved laborers were employed plowing and hoeing the corn fields for weeks on end, a task that according to Janney "kept us busy until mowing time." On Scott's farm the only break from this routine came on 7 June, when a number of hands were employed shearing sheep.

The first winter crops were ready to be harvested by the end of June, ushering in the most labor-intensive season in the agricultural calendar. The relatively small amount of rye was harvested first, but harvesting the vast amount of wheat consumed slaves' time. The grain was systematically scythed and cradled by field hands working in quick succession; then it was bound into sheaves. The binder "would take a bunch of wheat in his left hand, near the heads, with the right hand divide it in two parts, and by a dextrous turn, twist the two together so that when applied to the sheaf, it would hold fast." Finally, shocks of a dozen sheaves each were brought to the stackyard, and the stalks left in the fields were cut for hay to be used as fodder for the livestock. While the wheat and rye harvest lasted only until the middle or end of July, tasks such as binding, stacking, and cutting hay kept field hands busy throughout the summer. In the meantime, the oats were also harvested at the end of July or beginning of August, and a new crop of wheat was prepared for and planted in August and September. By then the annual cycle had begun anew.[7]

As Fairfax County planters adopted a system of diversified agriculture at the turn of the nineteenth century, they in fact turned old tobacco plantations into mixed farms. Their slaves became jacks-of-all-trades, skilled in a number of farm occupations from an early age. Local newspapers advertised the sale or hire of field hands who could perform any number of different duties. One typical advertisement read: "To Hire . . . a steady young NEGRO MAN, who has been accustomed to almost any kind of work." One local slave claimed that he was responsible for no less than forty-two different duties, "from mending roads and fences, to planting corn and shearing sheep."[8]

Moreover, as agricultural production dwindled in absolute terms during the course of the antebellum period, women on Fairfax County farms disproportionately found themselves working as domestic servants at some point in their lives. This was often far from an attractive alternative to farm labor, as domestic servants were at the beck and call of their masters both day and night and were made to perform an endless list of physically demanding chores. The experiences of the domestic servants at Wilton Hill, a plantation owned and run by John J. Frobel and his family, offer a case in point. The diary entries of John's daughter Anne mention that one servant named Milly had to cook for irritating hosts who repeatedly "inquired of [her] if dinner could not be an hour earlier than normal." She was also in charge of "milking the cows" at dawn, often sent on foot to carry messages to other farms throughout the neighbourhood, and made to clean and perform several other duties. Rose, another servant at Wilton Hill, became fed up with one of her mistress's northern guests, who constantly bothered her for "cold water, and warm water and hot water, and towel after towel, and soap—every sort of thing until Rose's patience is entirely exhausted." Rose finally blew up at the woman, yelling "Well Mrs. Yank, you may wait on yourself now for I won't put my foot in that room again while you are on the place."[9]

For slaves employed in agriculture, evidence indicates that planters and farmers in Fairfax County maintained a semblance of traditional work patterns from the old tobacco culture, even as they shifted production from tobacco to grains. Time-work remained the preferred system for most field labor, and field hands worked six days a week from sunup to sundown, sometimes alongside or under the direct supervision of their masters, either in small squads or singly. (The decline in slaveholding size resulted in markedly smaller squads in the nineteenth century than in the eighteenth, however, frequently consisting of only two or three hands.) To maximize efficiency, squads were usually interdependent and made responsible for specific tasks. Hoe squads followed the plows, for example. During the wheat harvest one squad would scythe and cradle, a second would rake and bind, and a third would come along and stack. The cultivation of grains thus very much required a team effort—individual tasks were limited to milking cows and tending livestock; cultivating the farm's vegetable garden; and irregular work such as repairing fences and shearing sheep. Even when working singly, however, slaves were kept busy from dawn to dusk, and they were often shifted in and out of squads in the grain fields.[10]

Wheat, corn, oats, and rye are by no means as fragile or fickle as tobacco, but several factors appear to have been responsible for the maintenance of the old work system in northern Virginia. First, ganging and time-work had already been applied to wheat cultivation during the tobacco era, when it was a secondary staple. Indeed, most planters probably did not even consider alternative work patterns in the nineteenth century. Second, the exhaustion of the soil and the economic decline that plagued local agriculture made close supervision of land and labor essential to reversing the downward spiral, and gang labor facilitated such supervision. Finally, time-work allowed planters to adequately meet the seasonal time constraints involved in diversified agriculture. As antebellum field hands in Fairfax County were employed in the cultivation of not one cash crop but (at least) four—all in quick succession—their operations were often characterized by haste. Philip Morgan found that Chesapeake grain producers organized their slaves into gangs in order to "drive them with greater intensity."[11]

Evidence suggests that the workload for many northern Virginia field hands actually increased in the nineteenth century. As slave labor forces dwindled, whether by sale or long-term hiring, fewer field hands became responsible for a greater amount of work, and these duties had to be performed quickly to meet seasonal time constraints. Historian Lorena Walsh found that hard-pressed planters and farmers throughout antebellum northern Virginia attempted to extract "more work from their slaves by increasing the length of the workday and the amount of off-season labor, as well as dividing more tasks by gender and age."[12]

In practice, time-work in Fairfax County burdened slaves with long workdays and very little free time or flexibility. The haste and physical exertion of their work were interrupted only by brief meal breaks at noon, and their free time was limited to the darkness of the evening hours and Sundays. Describing the work patterns of enslaved people in northern Virginia, historian Brenda Stevenson noted that "their work was hard and often monotonous, their workdays were long," and of leisure time there was "precious little." This appears to have indeed been the case for most. Austin Steward, a bondsman from the area, recalled in his autobiography that "it was the rule for the slaves to rise and be ready for their [work] by sun-rise, on the blowing of a horn or conch-shell," and that they were not permitted to retire to their cabins till dark. Christopher Nichols, another local bondsman who successfully escaped to Canada, related to interviewers: "We were up before day—when the rooster crowed, the horn blowed. By the time one could see his hand before him, he was at work, and we were kept at work until late." And George Jackson, also enslaved on a local wheat plantation, recalled that the slaves on his farm "all worked hard and late at night." Even visitors to the region commented on enslaved people's especially long workdays. Elijah Fletcher wrote during his stay at Hollin Hall in 1810 that the slaves there worked "frequently casting an eye to the sun to see if it is not noon [for their dinner break] and then to see if it is not night," as their workdays lasted from "sunrise till sunset."[13]

During especially labor-intensive seasons, such as the wheat harvest, field work often lasted from before dawn till long after sunset. The owner of Walney plantation claimed that during the harvest season he and his field hands "rise *before* four, come in to supper about nine, and give up to breakfast and dinner an hour and a half of the intermediate time." Francis Henderson, a bondsman from northern Virginia, told interviewers after his escape to Canada that slaves on his farm lacked the time to cook their own meals during the harvest. "In harvest time, the cooking is done at the great house, as the hands are wanted more in the field," he recalled. Other seasons were less intensive, however. During late December and January, as daylight hours were limited and chores less regular, enslaved people probably enjoyed a welcome, if brief, respite from especially long workdays.[14]

Tasking in Stagnant Water: Georgetown District, South Carolina

Plantation labor in the lowcountry differed markedly from farm work in northern Virginia. Contrary to their Fairfax County counterparts, slaves in Georgetown District did not cultivate continually alternating grains throughout the year—with the notable exception of provision staples, rice was the only crop that mattered. But its cultivation was doubtless one of the unhealthiest and most labor-intensive occupations in which enslaved field hands in the antebellum South engaged. Elizabeth Allston Pringle, the daughter of prominent Georgetown rice planter Robert F. W. Allston, certainly hinted at the exhausting nature of the crop's cultivation when she declared, a touch dramatically, that "like the pyramids, slave labor only could have accomplished it." Moreover, the rice cycle was unique in its straightforward nature; field labor was well defined and predictable. "Nothing is left to chance," as local planter J. Motte Alston recalled, a factor that significantly influenced the development of work patterns for the bondspeople who toiled year round in the fields.[15]

As was the case with wheat cultivation in northern Virginia, the rice cycle in Georgetown District began in autumn, immediately following the harvest.[16] First, a number of field hands were put in charge of burning the trash and stubble left by the previous year's crop, while others were employed in the dreaded "mudwork"—ditching, embanking, and repairing the trunks. Proper function of these—a feature unique to the rice country—was of crucial importance. The trunks were long wooden sluices situated along the riverbanks, equipped with hanging doors on each end that opened and closed automatically with the change of the tide to allow for either the flooding or drainage of the rice fields, whichever was called for. The trunks were opened during flows and closed when the fields needed to be drained again.

The winter months did not usher in a slow period in the rice cycle, as they did in Virginia. Quite the contrary. During winter the land had to be prepared for the next year's crop by breaking it up—"chopping, mashing and breaking," as the slaves called it—which was a physically exhausting occupation indeed. Although traditionally accomplished with the hoe, more and more planters chose to also employ horse- or ox-driven plows during the latter decades of the antebellum period. Robert F. W. Allston wrote in an essay on rice cultivation in the 1850s that on his plantations, "we begin preparation for a new crop by . . . plowing the land as soon after the harvest as the fields can be gleaned." In most cases, however, land was initially broken up with hoes, then plowed over when the land was dry enough.[17]

As spring approached, field hands were put to work with their hoes making trenches in which to sow the rice seeds, a task requiring considerable skill and precision. The trenches were made twelve to fifteen inches apart, two inches deep, and three to four inches wide. Some planters began to employ plows for this task by the end of the antebellum period. When former slave Ben Horry heard his younger wife tell interviewers in the 1930s that trenching was accomplished by an ox-driven "mis-sheen" [machine], he interjected: "Dat mis-sheen come in YOU day, darling! My day [when] I trenching, hoe trench dat!"[18]

By the end of March, beginning of April, the fields were ready for planting. Prior to planting, however, the rice seeds were specially prepared by wetting them in clay and water, and then allowing them to dry. This was meant to prevent the seed from floating to the surface during the first flow. Then, when the seeds were ready, they were sown by hand over the entire width of the trenches and covered by hoes or rakes. As soon as the seeds were covered, the trunks were opened and the rice fields were flooded for the first time, a phase in the rice cycle called the "sprout flow." The land remained covered in shallow water for three to six days, just long enough for the grain to sprout. During the sprout flow, field hands were employed on the provision grounds planting corn and sweet potatoes. When the rice seeds had sprouted, the water was drawn off and the young plants were exposed to the sun and allowed to grow. At this time, the fields were weeded and guarded carefully from birds.

When the plants had reached the height of six inches the fields were again inundated for three to six days during the "point flow." The water level was then gradually lowered and finally drained, and the fields were dried for hoeing. Field hands began with a light hoeing, but after the plants were strong enough they followed with a second hoeing to remove any weeds that might have sprouted in the meantime. When this task had been completed, the "long flow" was put on, during which the plants were completely covered with water. This flow served to float off trash and destroy insects. After three or four days the water level was lowered to about six inches and kept on for a period of two to three weeks. During this period, some field hands were employed on the provision grounds of the plantation, while others waded through the muddy rice fields, pulling weeds and grass that could be seen above the water level. When the water was drawn off gradually and the land was dried, slaves returned to the muddy fields for two more hoeings. This phase of dry culture was followed by a final flooding, the "lay-by flow," during which the work tempo for field hands was somewhat reduced. The water was kept on for seven to eight weeks this time, and slaves again carefully waded through the fields, up to their knees in mud and stagnant water, pulling by hand any weeds, grass, or "volunteer rice"—rice that had come up from the refuse of the previous year's crop. They also guarded the plants from rice birds.

By the end of August, beginning of September, the rice was fully ripe and ready to be harvested and processed into a marketable commodity, ushering in the most labor-intensive phase in the rice cycle—"a season . . . of laborious exertion," in the words of Robert F. W. Allston. The fields were drained from the lay-by flow, and the following morning field hands commenced cutting the grain with sickles known as rice hooks. The stalks were left a day on the stubble to dry, then tied and stacked on flatboats and transported to the threshing yards. Threshing rice entailed separating the heads from their stalks by beating the grain repeatedly with flailing sticks. One visitor to a local rice plantation observed "about twelve female slaves, from eighteen to twenty-eight years of age, threshing rice on a sort of clay floor. . . . It was extremely hot, and the employment seemed very laborious." Next the rice was winnowed, whereby the grain was separated from the chaff, and pounded with large wooden mortars and pestles in order to remove the outer husks and inner cuticles. Threshing and pounding were among the most physically exhausting tasks for enslaved field hands in Georgetown District, and they required considerable skill and energy.[19]

Most of the threshing and pounding was still performed by hand as late as the mid-nineteenth century. However, in the 1850s some of the wealthiest planters built water- and steam-powered rice mills on their plantations to increase output. At Chicora Wood, rice was pounded "in mills run by . . . water power," and its owner Robert Allston wrote that ideally rice should be "threshed out by one of Emmon's Patent Machines." By 1860 there were eleven rice mills in use in Georgetown District; planters who did not own their own mills usually paid for their rough grain rice to be threshed and pounded at one of their neighbors' mills. J. Motte Alston, for example, paid to have the rice from his estate Woodbourne pounded at Laurel Hill, a plantation owned by his neighbor and friend, Francis Marion Weston. Rice mills did not eliminate threshing and pounding by hand, however; often both methods were employed simultaneously. Former bondswoman Maggie Black remembered that as late as the Civil War, enslaved people still threshed rice by hand on her plantation: "Heap of people came from plantations all about and help whip dat rice." Likewise, she recalled that the traditional method of pounding was employed on her plantation: "Dey have rice mortars right dere on de plantation what dey fix de rice in just as nice." A northern visitor to Benjamin Dunkin's plantation also reported seeing field hands at work threshing and winnowing by hand while a water-powered rice mill was employed for pounding. Once pounded, the finished product was packed into wooden barrels and transported to Charleston, from where it was shipped to various domestic and foreign destinations.[20]

Just as cultivation methods on the plantations of South Carolina differed from those in other parts of the South, so did work patterns. Elsewhere most enslaved field hands performed time-work and toiled from sunup to sundown, but lowcountry slaves worked according to the task—or piece-work system, an organization of labor that developed there in the eighteenth century. Under this arrangement, for six days a week, field hands were assigned a fixed amount of work that they were required to complete by sundown the latest, because "if you didn't, then you got a whipping." (Half tasks were common on Saturdays.) When they had finished their tasks, the rest of the day was theirs to spend as they wished. The amount of time that lowcountry field hands spent engaged in agricultural labor was therefore not fixed, and the enslaved were afforded the relative luxury of determining their own work tempo, an oddity that continually caught the attention of visitors. Captain Basil Hall, a Scottish naval officer, observed that on the Georgetown rice plantations "a special task for each slave is . . . pointed out daily by the overseer." Frederick Law Olmsted of the *New York Daily Times* wrote during his stay on a rice plantation farther south in the Georgia lowcountry that "nearly all ordinary and regular work is performed *by tasks*: that is to say, each hand has his labor for the day marked out before him, and can take his own time to do it in."[21]

The rise of the task system in the eighteenth-century lowcountry, as opposed to other parts of North America or even the New World, has been attributed to a number of factors, most importantly to do with the specific nature of rice cultivation. First, rice was not a fragile crop and required little supervision, unlike other cash crops such as tobacco. On the contrary, rice was "a hardy plant, requiring a few relatively straightforward operations for its successful cultivation," according to Philip Morgan. The long hours and constant supervision needed for tobacco cultivation—factors that convinced northern Virginia planters to employ the time-work system—proved unnecessary on the rice plantations. Second, the predictable nature of rice cultivation allowed daily tasks to be easily and fairly calculated and measured. Drainage ditches conveniently carved up the fields into quarter- or half-acre units that could be assigned to able-bodied field hands and easily monitored and measured by overseers and drivers.[22]

Some historians have explained the rise of the task system in the low-country as a result of rice planters' temporary absenteeism during the summer months, when—in contrast to other parts of the South—the white population usually fled to the mountains, beaches, or cities to escape the malaria. From a distance, planters supposedly could not force their slaves to remain in the fields from sunup to sundown and therefore "conceded control over worktime in return for a generally accepted unit of output," in the words of Ira Berlin. Ample evidence indicates that summer absenteeism was indeed widespread in Georgetown District. Elizabeth W. Allston Pringle, who grew up on Chicora Wood plantation, claimed that "the planters removed their families from their beautiful homes the last week of May, and they never returned until the first week in November, by which time cold weather had come and the danger of malarial fever gone." J. Motte Alston, who planted on the Waccamaw Neck, recorded in his memoirs that "the rice planters of Waccamaw would spend their winters on their plantations and their summers on the sea-islands near by. . . . I built a comfortable cottage at the Southern point of Pawley's Island." And Ben Horry, a former slave on Joshua John Ward's Brookgreen plantation, told interviewers that every year his "Marse and Missus go to North Carolina mountains, Broad River section" during the summer months.[23]

The absenteeism argument attributes the development of fixed labor quotas on the rice plantations mostly to the bargaining power of the enslaved, who according to Berlin "conspired to preserve a portion of the day for their own use while meeting the planters' minimum work requirements." Judith Carney has suggested that enslaved Africans even brought the notion of a task-based organization of labor in rice cultivation with them to the lowcountry. These explanations may be problematic, however. Although they were indeed usually absent during the summer months, most rice planters left black drivers in charge of plantation management and they would have had the authority to force slaves to remain in the fields till dark. Moreover, as Philip Morgan has pointed out, long-term absenteeism did not lead to the widespread development of task work on Caribbean sugar plantations, where planters and overseers worked their slaves according to the gang labor system instead. Planter absenteeism and slave agency certainly influenced the nature of lowcountry tasking, but, as Morgan argues, such factors do not appear to have been crucial to the system's inception.[24]

Where slaves did demonstrate their bargaining power was in defending what they deemed to be fair tasks. By the antebellum period, local custom and tradition determined what constituted a fair day's work, and if planters or overseers attempted to increase tasks, enslaved men and women generally responded by either refusing to work or by running away in droves. One planter warned that "should any owner increase the work beyond what is customary, he subjects himself . . . to such discontent among his slaves [as] to make them of but little use to him." Olmsted learned that elsewhere in the lowcountry, "if [a task] should be systematically increased very much, there is danger of a general stampede to the 'swamp'—a danger the slave can always hold before his master's cupidity." He continued that "it is looked upon in this region as a proscriptive right of the negroes to have this incitement to diligence [i.e., fair tasks] offered them."[25]

Tasks were adapted to the age and ability of each laborer. Able-bodied slaves on the rice plantations were divided into four categories: quarter-task, half-task, three-quarter-task, and full-task hands. In most features of the rice cycle, full-task hands were in charge of cultivating one quarter-acre unit of land per day. Others were required, according to age and ability, to cultivate only a certain fraction of a full task. Ben Horry recalled that on his plantation "everybody except chillun had a task. If you was cleaning, mashing, and levelling the ground where the rice would be planted, your task was a half or a quarter of an acre a day, depending on how old you were." J. Motte Alston wrote in his memoirs that "for every kind of work there was a set task, and so, according to ability, there were full-task, half-, and quarter-task hands." Olmsted also commented on the work patterns of the enslaved during his tour through the lowcountry. "The field-hands," he observed, "are all divided into four classes, according to their physical capabilities; the children beginning as 'quarter-hands,' advancing to 'half-hands'; and, finally, when mature, and able-bodied, healthy and strong, to 'full-hands.'"[26]

In theory, slaves' tasks were meant to constitute a full day's work, but evidence suggests that most planters in fact intended for their slaves to finish their tasks early. On the plantation of Plowden C. Weston, a full task was defined as "as much work as the meanest full hand can do in nine hours, working industriously." Weston cautioned his overseer moreover that "this task is *never* to be increased, and no work is to be done over task except under the *most urgent necessity*.... No negro is to be put into a task which they cannot finish with tolerable ease." Indeed, he continued, "the hands [should] be encouraged to finish their tasks as early as possible, so as to have time for working for themselves." James R. Sparkman, another local planter, claimed that on his estate "the ordinary plantation task is easily accomplished, during the winter months in 8 to 9 hours and in the summer my people seldom exceed 10 hours labor per day.... Whenever the daily task is finished the balance of the day is appropriated to their own purposes."[27]

Just how "easily" these tasks were accomplished is open to debate. Nevertheless, planters were very well aware that most tasks would not take all day to complete, and they conceded to their enslaved field hands this incentive to work industriously. Domestic servants on the rice plantations—who worked for their masters both day and night—often even envied the field hands their opportunity to maximize their free time. One traveler to Georgetown District found that the field hands enjoyed clear advantages over the so-called "privileged" domestic servants, because the field hand's "labors have their definite limits, and he knows when they will terminate for the day." For that reason he found the "field hand to be more contented, even buoyant in his spirits, than the domestic slave."[28]

Plenty of evidence confirms that slaves often finished their tasks unusually early. One nineteenth-century visitor to Georgetown District observed that slaves' tasks "were commonly not beyond what moderate diligence would enable them to perform, and by the vigorous and industrious were sometimes accomplished by 2 or 3 o'clock in the afternoon." Adam Hodgson, another visitor, likewise recorded that "the Negroes usually go to work at sunrise, and finish the task assigned to them at three or four." David Doar, an antebellum rice planter on the Santee, claimed that "a good and industrious hand could get home by 3 or 4 P.M. ... and night work was never required." Captain Basil Hall remarked that the "assigned task [is] sometimes got over by two o'clock in the day, though this is rare, as the work generally lasts till four or five o'clock." This was not unusual for the lowcountry. Olmsted wrote that slaves' tasks in the Georgia lowcountry "certainly would not be considered excessively hard, by a Northern-laborer; and, in point of fact, the more industrious and active hands finish them often by two o'clock. I saw one or two leaving the field soon after one o'clock, several about two; and between three and four, I met a dozen women and several men coming home to their cabins, having finished their day's work."[29]

On J. Motte Alston's plantation, slaves sometimes completed two tasks in one day, in which case they earned a free day. "When two tasks were accomplished in one day by any hand," Alston wrote in his memoirs, "he was not expected to work the next, and these tasks were *never* increased." To be certain, few enslaved field hands probably ever accomplished this feat. Nevertheless, it may be safely concluded that on average, enslaved field hands in Georgetown District were more flexible and had more time than slaves living in those regions of the antebellum South where time-work was employed.[30]

Ironically, the enslaved agricultural laborers in Georgetown District with the least amount of free time were precisely those with the most privileged status: the drivers. The managerial assistants of the overseers, drivers on the lowcountry rice plantations were responsible for a number of duties, including calculating and assigning the tasks of other field hands, inspecting completed tasks, and generally monitoring and supervising the day-to-day functions of the plantation. The driver's workday began earlier than that of the other field hands, because he was the one responsible for waking them up and shuffling them onto the flatboats that would ferry them across to the rice fields every morning. Plowden C. Weston instructed his overseer that the driver should be in charge of "bringing out the people early in the morning. Every negro [is] to be on board the flat by sunrise. One driver is to go down to the flat early, the other to remain behind and bring on all the people with him." During the day, the driver was responsible for "the proper performance of tasks . . . and generally for the immediate inspection of such things as the overseer only generally superintends. . . . No negro is to leave his task until the driver has examined and approved it."[31]

Being in charge of inspecting the last tasks meant that the drivers were the last ones to leave the rice fields. And although many slaves certainly finished their tasks by the early afternoon, others no doubt worked more slowly and remained in the fields until sundown. Ben Horry recalled his mother once being beaten by the driver for not finishing her task by sundown: "I see gash SO LONG (measuring on fore-finger) in my Mama—my own Mama!" Ex-slave Gabe Lance told interviewers that "any slave . . . didn't done task, put 'em in barn and least cut they give 'em with lash been twenty-five to fifty." Albert Carolina, another former bondsman from Georgetown, related one particular beating to interviewers of the Federal Writers' Project: "Mausser gin (give) the woman a task. Didn't done it. Next day didn't done it. Saturday come, task time out! Driver! I tell the truth, you could hear those people, 'Murder! Murder!'" Employing the lash, drivers may have unleashed their frustrations on slower-working field hands, as these slaves prolonged quitting time. One former field hand from the lowcountry opined that the driver on a rice plantation was always at a disadvantage, "for he had no task-work and no time of his own, while the other slaves had the Evenings to themselves."[32]

With the Precision of Military Drill: St. James Parish, Louisiana

By the time Philo Tower visited southern Louisiana in the 1850s the sugar plantations had already developed quite a reputation for severe and unremitting labor, even by southern standards. As he made his way north from New Orleans along the Mississippi, Tower pitied the enslaved men and women he saw toiling in the endless fields that extended at perpendicular angles from both sides of the river. The slaves, he recorded, were "worked up, body and soul, to aggrandize these lordly, popish descendents of France and Spain." He added that "none of the states are so much dreaded by the slaves as Louisiana." Charlotte Brooks, a former slave who was sold from Virginia to toil on a Louisiana sugar plantation, confirmed this statement when she related to an interviewer after emancipation: "I tell you nobody knows the trouble we poor colored folks had to go through here in Louisiana. I had heard Louisiana was a hard place for black people, and I didn't want to come." Enslaved people's particular dread of the southern Louisiana sugar country appears to have been justified. Like their counterparts in the rice country, slaves in the sugar districts worked in an unhealthy environment, and like their counterparts in northern Virginia they worked with haste during most of the year—but workloads on the sugar plantations surpassed anything found in the rest of the South.[33]

For slaves living in St. James Parish, the annual planting cycle began almost immediately after the previous year's sugar had been made, usually in early January, offering field hands little respite from one season to the next. A one-week holiday between sugar seasons was the norm but was not always consistently granted. Valcour Aime's slaves were granted only a three-day holiday between the 1849 and the 1850 sugar seasons. After the last sugar had been made on 6 January 1850, he gave his field hands "three days of rest, that is, the 7th, 8th and 9th," but he "resumed plantation work on the 10th." After their short break, the slaves' first chore of the new year was to prepare for cane planting by thoroughly plowing the ground and opening up deep furrows, six feet apart and seven to nine inches deep, into which the seed cane would be planted. The seed cane was prepared the previous autumn, just before the harvest commenced.

This was accomplished by cutting the cane that was to be used for seed and storing it in mats covered with leaves and cane trash to protect it from winter freezes. Once the furrows had been opened up in January, the seed cane (which consisted of stalks rather than actual seeds) was removed from its winter storage, stripped of its top and flags—or "unshocked"— and cut into convenient lengths. It was then placed lengthwise into the furrow, covered to a depth of three to four inches, and left to sprout.[34]

Planting was a laborious and time-consuming task that monopolized the plantation's work force. Solomon Northup, enslaved on a northern Louisiana cotton plantation but regularly hired out to the sugar country farther south, recalled in his autobiography that "three gangs are employed in the operation." The first gang "unshocked" the stalks and carted them to the rows, the second gang laid the seed cane in the furrows, and the third gang followed with hoes to cover the stalks. Although planters in Louisiana were in the habit of planting sugar every third year, as explained in chapter 1, this did not mean that slaves were spared the job of planting during two out of every three years. Sugar plantations in St. James Parish, as elsewhere, were divided into several sections, each planted in rotation during a different year. Planting was thus an annual task for slaves in the sugar country, despite the perennial nature of the crop.[35]

Once the seed cane was in the ground and covered, usually by the end of February, a couple of weeks were devoted to performing other kinds of work on the plantation. One of the most urgent of these duties was—as on the rice plantations—clearing out the drainage ditches that had "become choked up by vegetation in the course of the summer and fall months," according to one visitor. T. B. Thorpe reported during his visit to the parish that "the ditches form one of the most important and expensive necessities of a sugar estate; for, with the exception of frost, standing water is the most destructive thing to cane." While the ditches were being cleared, other field hands were employed burning the cane trash from the previous year's crop, building and repairing the levees, making bricks for construction, clearing more land, draining swampland, or chopping wood for fuel. Still others were set to "scraping" some of the dirt from the rows of planted cane to give the sprouts a better chance.[36]

When the first sprouts appeared in March, the cultivation of sugarcane began in earnest. All able-bodied hands were set to plowing and hoeing the rows of sugarcane, harrowing the middles between the rows to keep the weeds down. This process was repeated an average of five or more times at intervals of about two weeks. Cane needed to be protected from more than just weeds, however. In March, frosts still posed a potential threat to the young plants. Valcour Aime complained in his plantation diary on 24 March 1850 that the "canes in the rear of the plantation [are] slightly frost bitten," a disastrous situation that could result in substantial loss. When frosts threatened, earth was placed over the roots of each stalk to prevent freezing. During the spring and summer, sporadic torrential downpours and recurring droughts likewise required field hands to be on the alert for crop damage. Soil that had become trodden down and compacted by heavy rain needed to be loosened up with the plow. If the furrows became flooded, they needed to be dug deeper. When droughts occurred, the fields needed to be irrigated. As T. B. Thorpe noted: "From the time the cane is put in the ground it is the source of constant anxiety," a sentiment that contrasted markedly from the rice districts of the lowcountry. If all went well, cane was usually ready to be laid by in June or July. The rows were then ridged up to allow for proper drainage, and the cane was left to grow untended until harvest time.[37]

In between sessions of plowing and hoeing, and during the period after the cane had been laid by, enslaved field hands in St. James Parish were overwhelmed with an endless list of other plantation chores. Throughout the spring and summer, most planters regularly diverted their labor force to plant and cultivate one or two crops of corn, as well as additional provision crops such as potatoes, pumpkins, and other vegetables. Chopping and collecting wood for fuel was an endless task, especially in preparation for the sugar harvest. Some three to four cords of wood were required to produce one hogshead of sugar, so the demand for wood throughout the parish was insatiable. Thorpe reported in 1853 that "the primitive forests are rapidly disappearing before this consumption." On top of all that, slaves were expected to dig drainage ditches, repair levees, make bricks, mend roads and fences, and perform a number of other duties when they were not in the cane fields. In the suffocating heat and humidity of the Louisiana summer, such work demanded severe physical exertion of field hands. Valcour Aime reported in July of 1844 that the heat was so intense that it killed one of his best plow horses and nearly killed three others. One can only imagine how his two hundred plus slaves fared under such circumstances.[38]

The nature of the Louisiana cane harvest, or "grinding season," which commenced just after the corn harvest in October, was unique to the sugar country, ushering in a solid two- to three-month block of arduous round-the-clock shifts for the slaves. Like wheat in Virginia and rice in South Carolina, sugarcane needed to be processed into a marketable commodity;

cane, however, was in another category altogether. Virtually all cane plantations had their own sugar house, and endless rows of cylinder smokestacks marked their presence all along the riverbanks of antebellum St. James Parish. Unlike wheat or rice, cane begins to sour almost immediately after it is cut; therefore, planters could not afford to harvest their crop and wait to process it at a neighbor's mill during the winter months, as they often did in Fairfax County and Georgetown District. And if they had had to leave their cane standing in the fields until a neighbor's mill became available it surely would have been ruined by autumn frosts. Fully equipped with the necessary high-tech machinery to process cane into granulated sugar and molasses, sugar houses more closely resembled small factories than barns or sheds, especially after the 1830s when steam came to replace horse power. In the 1850s, one visitor reported that the machinery of a local sugar house consisted of a "steam-engine of eighty horse-power, and sugar-mill for grinding cane, engines, vacuum pans, and a complete apparatus for making and refining twenty-five thousand pounds of sugar every twenty-four hours direct from the cane-juice." Once these mills and engines were fired up and the harvesting of the crop began, the sugar making continued largely uninterrupted—twenty-four hours a day, seven days a week—until completed. Only when emergency repairs were needed on certain machinery in the sugar house or when rain delayed cane cutting did the furious work tempo of the grinding season even slightly subside.[39]

While preparations were being made at the sugar house, slaves began to harvest the standing cane. The initial priority for field hands was to select the best quality cane to be used for seed during the next planting. This was cut first and stored in mats as described above. After that, the precarious cane harvest began in earnest. Determining the proper time to commence cutting was no simple task, but once the decision had been made to begin, work was carried on with unbelievable haste to prevent spoilage—the cane had to be cut, carted to the sugar house, and boiled down in quick succession. "From the moment the first blow is struck," reported T. B. Thorpe, "every thing is inspired by energy." Armed with large knives, slaves entered the vast harvest fields in groups and began the backbreaking task of cane cutting. First they sheared the flags from the stalk, next they severed the top of the stalk, and finally they cut the stalk down at the root. The field hands who were not engaged in actually cutting the cane followed close behind to collect the stalks and toss them into carts for quick transport to the sugar house. The tempo was kept constant: cane cutting was timed so as not to overload the mills while keeping the supply steady enough not to delay operations. If the operation was overtaken by cold weather, the field hands were set to windrowing the standing cane so that it would not be ruined by frost.[40]

Once at the sugar house, the cane was processed into granulated sugar and molasses in an industrial fashion. The stalks were immediately placed on a leather conveyor belt and transported to a large grinding mill containing two or three heavy iron rollers. During the opening decades of the nineteenth century these mills were still powered by horses, mules, or oxen. In the 1830s, however, steam power was introduced, increasing output and productivity. The iron rollers crushed the stalks and squeezed out the cane juice. The extracted juice then fell into a conductor underneath the mill, from which it was carried into a reservoir and further conveyed by pipes into large filters. Here the juice was clarified by boiling, and slaves ladled off the impurities, which floated to the top. After the cane juice was properly purified, the next task was to reduce it into syrup by evaporation. The juice was carried by pipes from the filters to large open kettles or vacuum pans, where it was heated to the boiling point. The boiling juice then passed through three pans in succession, each of which further strained the juice and required impurities to be removed by hand. When the syrup reached the point of granulation it was poured into the "coolers," or troughs from ten to twelve feet long and one and a half feet deep. Once cooled and properly granulated, the sugar was removed to the "purgery"—a large room with a cement cistern instead of a floor, on top of which were laid strong timbers like floor beams. Empty barrels with filtering holes in the bottom rested on top of these timbers. The granulated sugar was poured into the barrels and left there so the molasses could filter out and drain into the cistern below. The sugar was then packed into hogsheads, and the molasses in the cistern was poured into barrels. All in all, the harvest and manufacture of sugar lasted two to three months, at least until Christmas if not the beginning of January, when the planting cycle repeated itself.[41]

The forced nature of sugar cultivation in southern Louisiana, where the growing season was seriously limited by weather constraints, was the most important factor in determining the work patterns of slaves. As historian Richard Follett aptly put it, "every planter in the cane world faced a race against time." Indeed, more than any other southern crop, sugar was successfully cultivated only by working with a furious haste that easily surpassed the steady tempo necessary to cultivate continually alternating grains in northern Virginia. On sugar plantations there never seemed to be enough hours in the day, and the "sugar slaves worked at a killing pace," in the words of Ira Berlin. Slack periods were for all intents and purposes nonexistent throughout the year, as the Louisiana sugar cycle lasted a full twelve months, during all phases of which time was of the essence. Moreover, the secondary chores on the sugar plantations were endless, from cleaning ditches to cultivating corn to chopping thousands of cords of wood for the steam engines and boilers in the sugar house.[42]

Unwilling and unable to leave any of their operations to chance, the sugar planters became masters in plantation management; they demanded absolute speed, the highest efficiency, and an industrial discipline of their enslaved labor forces. While task work, which shortened slaves' workdays in the rice districts of South Carolina, was not unknown in the sugar country, it was reserved only for paid overwork (such as chopping wood) and skilled work. Ordinary hands endured systematic time-work, gang labor in the true sense of the term. The reasons were threefold. First, sugar cultivation required careful supervision, much like tobacco in eighteenth-century northern Virginia. Second, operations on the sugar plantations were unpredictable due to the crop's vulnerability to the Louisiana climate. And most important, there was simply too much work to be done on most sugar plantations. Gang labor met planters' needs with respect to each of these concerns. Only during the grinding season was gang labor replaced by assembly line shifts in the sugar house.[43]

For virtually all field work, slaves were divided into interdependent squads and driven at a quick tempo from sunup to sundown, six days a week, under the watchful eyes of overseers and drivers. Planting cane, for example, was accomplished by three gangs—the first gang stripped the stalks, the second laid the stalks carefully in the furrows, and the third gang came along and covered the stalks with earth. Cane cultivation, which consisted of repeatedly plowing and hoeing the cane fields, was likewise accomplished by gangs. The plow gangs skilfully drove the oxen and plows in perfectly straight lines along the rows of cane, while the hoe gangs—generally much larger in number—followed close behind, loosening the earth, killing weeds, and harrowing in between the rows. In autumn, when the crop was ready to be harvested, field hands were again divided into "cutter" gangs that cut the stalks; "loader" gangs that loaded the stalks on the carts; and "hauler" gangs that hauled the carts to the sugar house. Even secondary tasks were performed by gangs. One visitor reported that after the cane crop had been laid by, "the negroes are divided into 'gangs,' some to be employed in gathering 'fodder,' some to secure the crop of corn . . . some to manufacture bricks, while the sturdier hands are busily employed in cutting wood."[44]

Whatever the job at hand, the employment of gang labor, combined with the furious pace at which these gangs were driven, made field work on the sugar plantations proceed with military-like precision, a feature that did not go unnoticed by travelers to the region. Amos Parker observed during the harvest season "armies of negroes in the fields," working with a discipline unmatched anywhere else in the slave South. J. W. Dorr, during his visit to St. James Parish in 1860, reported regularly seeing "long ranks of fifty to a hundred negroes, hoe in hand, working across the fields with almost the precision of military drill." And Solon Robinson thought the cane-cutting gang on one plantation "quite a uniform company, that might do the state some service in times of peril."[45]

For the slaves, such regimentation was backbreaking. One former slave, Daffney Johnson, told interviewers of the Federal Writers' Project: "[They] sho' did push us out in dem sugar fields." The exhausting work tempo was exacerbated by the set long hours. Pam Drake wrote during a visit to Louisiana in the 1850s that the slaves on the sugar plantations all worked "from light in the morning until darkness at night." James Conner, a local slave who fled Louisiana in 1857, told abolitionist William Still that he and his fellow bondsmen "commenced work in the morning, when they could just barely see; they quit work in the evening when they could see no longer." Rebecca Fletcher, enslaved in her youth on a sugar plantation, remembered that "slaves had to go to field before daybreak and didn't come home till after dark." And Ceceil George, also a former slave, told interviewers that on the sugar plantations, "you has to put your candle out early and shut yourself up, den get up while it's still dark and start to work." As in Georgetown District, the slave drivers were especially disadvantaged with regard to free time, as they were made to perform extra duties after the day's work had been completed. The overseer on William Minor's sugar plantation in neighboring Ascension Parish—a typical estate for the region—was instructed to require the drivers to "examine the [slave] quarters (after ringing of the bell at night to see whether the negroes are all at home or not) . . . every night."[46]

The work week was even often extended to include Sunday. Despite claims by the daughter of one former sugar planter that she "knew of only one planter who made his negroes work on Sundays," evidence suggests that work on the Sabbath was in fact nothing unusual in St. James Parish or anywhere else in the sugar country. One local planter recorded in his plantation journal on Sunday, 5 September 1852: "All hands worked on pond till 9. Finished it." Ceceil George claimed that during slavery, "Sunday, Monday, it all de same . . . It [was] like a heathen part of de country." Elizabeth Ross Hite, another former bondswoman, likewise recalled that "de slave worked on Sunday and Saturday, sometimes," but added that "Sunday work was light." Even Valcour Aime, considered a model sugar planter by his peers, called his slaves to the fields on Sundays when he felt it necessary. In early July of 1844, he recorded in his plantation diary: "The fields being so unusually grassy, the hands were employed to weed on Sunday, the 7th."[47]

The grinding season combined the agricultural model of gang labor with modern methods of industrial production, a system that "emulated the emerging factory system," according to historian Richard Follett. The production of granulated sugar and molasses, once commenced, continued all day and all night for two to three months on end. Sugar planters divided their slaves into shifts to operate the machinery and perform the various functions of sugar production, with individuals often clocking eighteen hours of work a day. In most cases, able-bodied slaves were required to work two nine-hour shifts, with a six-hour break for sleep in between. The most valuable and skilled slaves, such as the sugar makers, however, could seldom afford to leave their posts at all and often slept in the sugar house.[48]

William Howard Russell of the *Times* wrote during a visit to St. James Parish that "when the crushing and boiling are going on the labour is intensely trying, and the hands work in gangs night and day." Another visitor reported that "the negroes . . . work from eighteen to twenty hours" during the grinding season. And Olmsted was appalled to find that "on all sugar plantations . . . they work the negroes excessively, in the grinding season; often cruelly. Under the usual system, to keep the fires burning, and the works constantly supplied, eighteen hours' work was required of every negro, in twenty-four—leaving but six for rest." Despite long hours, most slaves were enticed to the sugar house by certain incentives such as more and better food and drink, and planters did their best to turn the grinding season into a festive occasion. For those who stuck it out, work dominated their waking hours.[49]

Indeed, whether toiling in the fields or working night shifts at the sugar house, the excessive labor required of enslaved people throughout the year in southern Louisiana was gruelling even by southern standards. Visitors were often appalled by what they found to be the most inhumane work patterns in the entire plantation South. British traveler Thomas Hamilton voiced a common sentiment when he wrote that the slaves living on sugar plantations were all "compelled to undergo incessant labour . . . taxed beyond their strength, and . . . goaded to labour until nature absolutely sinks under the effort." James Stirling, another visitor to the sugar country, opined that "the severe nature of the labour on a sugar plantation . . . is objectionable both in an economical and social point of view." And Pam Drake found that throughout the sugar country, "the slaves are over-worked, at least one-third beyond what it is possible for the physical powers of any human being to endure."[50]

Conclusion

The work patterns forced upon enslaved people living in various regions of the nineteenth-century South differed widely. In Fairfax County, Virginia, the legacy of tobacco combined with the steady tempo required to cultivate continually alternating grains convinced local slaveholders of the need to work their bondsmen according to a slightly altered form of the gang system. As a result, slaves in Fairfax were made to perform time-work in small interdependent squads from sunup to sundown, cultivating wheat, corn, rye, and oats in quick succession. The cultivation of rice in Georgetown District, South Carolina, placed very different demands on enslaved field hands. There the hardy nature of the cash crop, as well as its relatively straightforward methods of cultivation, resulted in a unique system of labor. Working by the task, slaves on the rice plantations were able to determine their own work tempo, frequently avoiding the dawn-to-dusk labor of their northern Virginia counterparts. The work patterns employed on the sugar plantations in St. James Parish, Louisiana, contrasted greatly with those in the rice country and northern Virginia. Sugar cultivation in Louisiana was characterized by a severe lack of time, and the workloads on most plantations were substantial. The speed, precision, and long hours required in planting, cultivating, and harvesting sugar—not to mention performing a myriad of other tasks—convinced sugar planters of the need to employ a form of gang labor as it existed nowhere else on the North American continent. Moreover, slaves in St. James Parish experienced arduous eighteen-hour workdays during the grinding season.

The work patterns of enslaved people throughout the non-cotton South had far-reaching consequences for family contact and the development of family-based internal economies, as the following chapters will show.

3

Family Contact during Working Hours

The nature of work on agricultural units in various localities of the non-cotton South and the development of regional agriculture in general, examined earlier, set the context for the following two chapters; these focus on the effects of work and agriculture on the daily experiences of the men, women, and children who lived in bondage in each of the three regions. Enslaved families encountered a complicated framework of boundaries and opportunities in both their public world of work for the master and their somewhat more private world of the slave quarters. For this study it is convenient to divide slaves' waking hours into two parts: their time for the master, which consumed most of their day, and their time for themselves, including evenings and free days such as Sunday.

This chapter addresses the experiences of slave families during working hours. How did work and the nature of regional agriculture affect slave family contact during their time for the master? What kinds of boundaries and opportunities did work patterns and the specific demands for cultivating various cash crops create for parenthood and child care during working hours? Under what circumstances were enslaved people afforded the opportunity to work together with their family members during the day? And how did slave families react to these boundaries and opportunities? The answers to these questions are further explored here for slave families living in northern Virginia, lowcountry South Carolina, and southern Louisiana.

A Lack of Bonding Time: Fairfax County, Virginia

Labor arrangements and the nature of slave-based agriculture in northern Virginia conspired to limit slaves' control over their working hours and severely restrict opportunities to negotiate for family contact. As in other regions of the slave South, families in Fairfax County were generally confronted with systematic segregation during the workday. Such divisions ran two ways. Family members were separated both by age—young, able-bodied, and elderly—as well as by sex, according to their perceived ability and usefulness on the farm. The extreme lack of flexibility for slave families in Fairfax County in negotiating around such segregation was particularly trying, however, especially for parents with small children. As the following will show, work and the nature of local agriculture in many ways thwarted opportunities for family contact on the grain farms; however, contact between some able-bodied family members in the fields—especially on small farms—may have been common and probably improved over time.

Pregnancy and child care in Fairfax County posed an especially difficult challenge that disproportionately burdened enslaved women with the responsibility of reconciling their formal work with parental duties. Faced with declining productivity and shrinking labor forces, slaveholders in northern Virginia required physical labor from all women whether they were pregnant or not; consequently, pregnant bondswomen and the mothers of newborn infants frequently found that the pressing demands of mixed farming overrode the needs of their unborn or newborn children. Lacking both time and flexibility, local enslaved women found this a situation they were usually powerless to change.[1]

In northern Virginia, pregnant women were kept in the fields until shortly before they gave birth, although advanced pregnancy usually meant less work, if not in theory then in practice. One local overseer complained to his employer that "2 or 3 of the Negroe women are pregnant which will throw me behind in my crops," indicating that these women worked more slowly than the others, perhaps on purpose. Even when they were near their time and unable to perform some of the more strenuous field tasks, however, they could always be given lighter chores around the farm until they gave birth, from milking cows to weeding in the vegetable garden.[2]

Quite soon after their children were born, moreover, new mothers in northern Virginia were expected to fulfill their normal labor quotas again. For slave women in the antebellum South as a whole, according to historian Sally McMillen, four weeks appears to have been the average confinement period, or "lying-in period," after childbirth. Yet research by Brenda Stevenson has revealed that slaveholders throughout northern Virginia permitted an average lying-in period of only about two weeks before ordering new mothers back to work (a practice that also appears to have been common in the rest of the state according to historian Wilma Dunaway). Enslaved women in the region had little choice in the matter, and indeed were often even called upon to perform odd chores and favors during their two-week lying-in period. Former slave Christopher Nichols, who fled to Canada in the 1850s, recalled that on his farm a slave woman named Mary Montgomery, who had recently given birth and "had a small child at her breast," was made to do favors for her master during her confinement period, once even being sent out to get some ice, whereupon she caught a cold and became sick for days thereafter.[3]

Local slaveholders may have deviated from the southern norm when they limited slave women's confinement periods to the bare minimum, but in northern Virginia they could generally afford to work their childbearing women as hard as they did. Heavy labor quotas for pregnant women and short lying-in periods do not appear to have significantly raised mortality rates among women of childbearing age (which in 1820, for example, was only 3 percent higher than mortality rates among slave men), nor did they negatively impact child mortality rates. In 1820, the mortality rate for children under the age of fourteen in Fairfax County was 45 percent, even slightly lower than mortality rates among local white children. By 1850, mortality rates among slave children had decreased to 40 percent. As northern Virginia was not a diseased environment, and farm work was far from a deadly occupation, slaveholders were confident enough of their slaves' health to keep expecting women in the fields even during advanced stages of pregnancy, and most did not think twice about putting new mothers back to work after what even by nineteenth-century standards amounted to a relatively short rest. In this they differed from slaveholders in less healthy regions (such as the lowcountry or southern Louisiana).[4]

After their brief lying-in period, enslaved women in the region were also denied the flexibility to return to the quarters at regular intervals during the day to nurse their newborn infants or spend any time with them. Indeed, most new mothers were afforded the opportunity to return to the quarters only during their midday meal break, which was the only time when any of the slaves' work was temporarily halted according to Elijah Fletcher, the tutor employed at Hollin Hall. Visitors to the South who frequently commented on specially granted nursing visits for slave mothers in various other southern localities failed to observe any such privileges in Fairfax County. Lacking opportunities to alter their situation in the balancing act between motherhood and slavery, improvisation became a valuable skill in child care on farms throughout the region. Sources from other Virginia slaves have shown for example that slave women sometimes crafted substitute "nipples" in order to keep their babies calm while they were out working in the fields. One former bondswoman recalled that mothers used to "keep de babies from hollerin' by tying a string 'roun' a piece of skin an' stickin' it in dey mouth.... I b'lieve dat meat-skin-suckin' help babies." Still, limited nursing deprived new mothers of valuable bonding time with their newborn children.[5]

Indeed, the experiences of slave families in Fairfax County were characterized by a lack of bonding time. For enslaved parents, child care was predominantly a shared responsibility in northern Virginia, as dawn-to-dusk labor allowed parents little time to attend to their children during the day. Many women did not even have enough time to prepare meals for their children early in the morning or in the evenings. Former bondsman Francis Henderson remembered that "the overseer's horn would sound" every morning before most women got a chance to "bake hoe cakes" for themselves or their children. Consequently, he claimed, he was often simply given a piece of stale bread or a herring to eat as his mother made haste to the fields. "I never sat down at a table to eat ... all the time I was a slave," he later recalled. Brenda Stevenson found in her study of slave family life in northern Virginia that "given the work loads of slaves [in the region], one or two persons, even the child's parents, rarely were able to attend to all of the components of the child-rearing task. Parents, therefore, had to share childcare and -rearing with others." Enslaved parents' dependency on others to raise their children was for many a hard pill to swallow. Caroline Hunter was born a slave in southern Virginia, but the remarks she made concerning parenthood were just as applicable to the northern part of the state. "During slavery," she recalled, "it seemed lak yo' chillun b'long to ev'ybody but you."[6]

Whenever possible, substitute child rearers were drawn from resident family members—extended or nuclear—who were either too old or too young to perform productive labor for their master. In most cases these caretakers were women, usually elderly grandmothers or aunts of the children in their care. One former bondsman recalled that as a child he was kept "in the house with my aunt" during working hours, his aunt being "quite an old woman." Laura, a slave woman belonging to M. C. Fitzhugh in the 1850s, probably left her infant daughter under the care of her grandmother, Truelove, while she was in the fields. At seventy-five, Truelove was valued at nothing, and she lived in the same house as her granddaughter and newborn great-granddaughter. The significant contribution of elderly family members to child rearing reinforced kinship ties, as young enslaved children were socialized by authority figures other than their mothers and fathers. Frank Bell was not cared for by his grandmother during the day, but he did remember that his grandfather advised him on important aspects of life in bondage. During an interview in the 1930s he recalled how his grandfather had once called him over as a small boy to explain to him the nature of life as a slave: "'Come here, son,' he say to me, and it was the fust time I recollec' dat he had say something to me what wus a little tot dat ain't nobody paid 'tention to. He picked me up and rid me cross his foot wid his knee crossed, holding onto my hands, and riding me up and down. 'Son,' he say, 'I sho' hope you never have to go through the things your ole grandpa done bin through.'"[7]

Calculations derived from the federal census returns for the antebellum period, however, indicate that on average only about one-third of enslaved children living in Fairfax County could have in fact been cared for during the day by elderly women who lived on the same slaveholding. Since local slaveholdings were generally small, many farms simply did not have any elderly enslaved women who could look after small children. The census returns for 1840, for example, indicate that of the 1,170 slave children living in Fairfax County, only 399 resided on estates that also contained at least one enslaved woman aged fifty-five or older. In practice the number of children cared for by elderly slave women was undoubtedly even lower, as many female slaves were kept in the fields past the age of fifty-five, while others were transferred to the big house to perform domestic duties when they became too old to perform field work. In other words, in the *best case scenario* only 34 percent of Fairfax County slave children could have been supervised by resident elderly women—whether family members or otherwise—while their parents were in the fields. The plantation of David Wilson Scott was fairly typical in this regard. In 1820 he owned eleven slave children but no slave women above the age of forty-five. Of the four women that he owned between the ages of fourteen and forty-five, two were employed as field hands. The other two worked as domestic servants.[8]

Clearly most enslaved mothers and fathers had little choice but to share their child-rearing tasks with other residents of the farm. Their options were limited. On small farms throughout the South it was common for young children to be put in charge of looking after their infant brothers and sisters while their parents were away, although sources for Fairfax County or northern Virginia do not indicate how widespread this practice was. Evidence suggests that many enslaved boys and girls past infancy in fact fell under the loose supervision of domestic servants (in some cases their own mothers or extended family members) or even the white mistress, as black children were often observed running around the big house and playing with the white children. Former bondsman Christopher Nichols recalled that on his farm, the black and white children played together in the "stackyard" during the day. The son of one northern Virginia slaveholder likewise claimed that it was customary "in nearly all households ... for the white and black children connected with each to play together." Frederick Law Olmsted noted with some surprise during a railway journey through the region in the 1850s that "black and white children are playing together . . . black and white faces are constantly thrust together out of the doors, to see the train go by." Bethany Veney, born a slave in northern Virginia just west of Fairfax, claimed that she had been supervised by her white mistress, a certain Nasenath Fletcher, while her mother worked in the fields. Fletcher apparently even took it upon herself to teach the impressionable enslaved children under her care that in the next life, "every little child that had told a lie, would be cast into a lake of fire and brimstone." In her autobiography, Bethany recounted asking her mother if what her mistress had told her was true. No doubt resentful of the lessons being taught to her daughter by the white woman, Bethany's mother "confirmed it all, but added what Miss Nasenath had failed to mention, namely that those who told the truth and were good would always have everything they should want."[9]

Some sources indicate that enslaved children were sometimes left completely unsupervised while their parents were away. Slave parents living on small holdings—and local slaves increasingly found themselves living on small holdings as the nineteenth century progressed—often had no other choice. Charles Peyton Lucas, a former bondsman from the area, told interviewers in Canada (after his escape from Virginia) that he was simply "kept mostly in the quarters" as a young boy while his parents were away. Even slave children on moderate-sized holdings were not always formally supervised. Fairfax County slaveholder Richard Marshall Scott, Sr., owned thirteen slave children under the age of fourteen and seven slave women above the age of forty-five in 1820, giving the impression that he had at least a few potential children's nurses in his possession. However, as he divided his slaves between two farms and his town house in Alexandria, it appears that at least some children ended up on units where they were left unsupervised while their mothers were away. Scott received an unwelcome surprise one afternoon by the unsupervised children of one of his enslaved women. On 8 April 1823 he noted in his diary with disdain: "On my return from Alexandria I found my servant hall, my barn, stables, carriage house, covered way, barnyard, hay, straw, plow, cartwheels . . . [and] 13 cords of wood consumed by fire, which broke out in the servants hall on Sunday the 6th. . . . The fire was occasioned in the absence of a mother by her little children setting fire to a straw bed." In this particular case the children's mother was not in the fields—it was Sunday—but rather in Alexandria, where she may have been temporarily serving in Scott's town house. Nevertheless, the incident indicates that it was not unusual for children to be left unsupervised on the farm.[10]

Partly as a precaution to prevent such behavior, and partly as a means of utilizing their enslaved labor force to its maximum potential, Fairfax County slaveholders often put otherwise unsupervised children to work at a relatively young age. Many ended up serving as domestic servants in the big house until they were old enough to perform field labor. From the perspective of the parents, slave children's domestic duties meant, in the words of Brenda Stevenson, "working long hours, sometimes being accessible twenty-four hours a day," and thus further limiting valuable family time. When northern journalist James Redpath called at a small grain farm in Fairfax County in 1857, the door was opened to him by "a young negro girl, six or seven years of age," who served her mistress in the house. Austin Steward was taken away from his mother, father, and sister at the age of eight to serve as his master's personal servant. "It was my duty to stand behind my master's chair," he recalled, "which was sometimes the whole day, never being allowed to sit in his presence." Steward was even prevented from sleeping with his own family, as his master and mistress required him to sleep in their room in case they needed him during the night: "I slept in the same room with my master and mistress. . . . I always slept on the floor, without a pillow or even a blanket, but, like a dog, lay down anywhere I could find a place." Another Fairfax County slave named Oscar was taken into the big house at a young age to serve as his mistress's "pet." A report filed with abolitionists in the northern city to which he fled stated that his mistress, one Miss Gordon, "raised Oscar from a child and treated him as a pet. When he was a little 'shaver' seven or eight years of age, she made it [a] practice to have him sleep with her" instead of with his own mother.[11]

Other young children were put to work outside, performing light tasks such as collecting trash or stones about the estate, or toting water to the fields where the adult field hands were working. Most children began to work when they were about six years old. Former bondsman George Jackson remembered working in the farm's communal vegetable garden, under the supervision of the white mistress, when he was five or six: "I worked in the garden, hoein' weeds. . . . De mistress scold and beat me when I was pullin' weeds. Sometimes I pulled a cabbage stead of a weed. She would jump and beat me. I can remember cryin.'" Some children were sent to help the other field hands—in many cases their own mothers, fathers, or older siblings—to bind and stack wheat, pull weeds, or remove worms and other pests from the crops. Once a child was deemed old and strong enough, he or she was made to perform regular work on the farm. The exact age for this transition varied according to the opinion of the slaveholder; it could occur as early as six or as late as fourteen years old, but the average age seems to have been between eight and nine. Silas Jackson, a former slave from northern Virginia, was nine when he began to perform regular work. He remembered that he was "a large boy for my age. When I was nine years of age my tasks began and continued until 1864." Henry Banks, also a former bondsman, was even younger. "When I was eight years old," he claimed, "I was put to work regularly on the farm."[12]

The extent to which Fairfax County field hands were afforded opportunities to work in family groups is difficult to determine. Very little primary evidence exists on the subject; consequently, conclusions can only be derived from a limited number of sources. Suffice it to say that the advantages of such an arrangement for slave families would have been substantial, however. Even if most slave families in the region lacked sufficient free time to be together or care for their children during the day, working in family groups at least would have allowed family members of working age to be together and even assist one another as they labored in the wheat and corn fields. Although squads of complete nuclear families in the fields were undoubtedly an uncommon sight in Fairfax County—as most slave families were at least partially scattered across the landscape on different slaveholdings (see chapters 6 and 7)—cooperation between some resident family members may have been common.

On a handful of the larger plantations it appears that slave families were granted the privilege of working together. Austin Steward recalled that "it was usual for men and women to work side by side on our plantation." Frank Bell, reflecting on his childhood in bondage, told interviewers in the 1930s: "We used to wuk in family groups, we did. Now me and my four brothers, never had no sisters, used to follow my mom an' dad. In dat way one could help de other when dey got behind." As Bell's testimony makes clear, slaves who worked in family groups were especially eager to come to the aid of family members while performing their master's work. Indeed, by helping family members when they "got behind," enslaved people essentially protected their loved ones from the lash or other punishments that were often meted out when their work tempo was found wanting. One former bondsman claimed that his master's son would sadistically sit "on the fence looking down upon us, and if any had been idle, he would visit him with blows. I have known him to kick my aunt . . . and I have seen him punish my sisters awfully with hickories from the woods." Men were especially keen to protect female family members. As Bell remembered it, "All of us would pitch in and help Momma who warn't very strong. Course in dat way de man what was doin' the cradlin' would always go no faster dan de woman, who was most times his wife, [so that she] could keep up." As enslaved men in theory lacked the power to protect their female family members from corporal punishments meted out by whites in positions of authority, those working in family groups seized opportunities to at least prevent their wives, daughters, and sisters from being physically abused by initiating slowdowns, or by personally assisting them with their work.[13]

On Bell's plantation, the privilege of working in family groups was negotiated by the driver. As in the lowcountry, drivers in Virginia held a managerial position within the plantation division of labor; the driver's duty was to set the pace of the rest of the gang through his own labor, generally oversee the agricultural operations performed on the farm, and even mete out punishments to his fellow bondsmen if necessary. Ex-slave interviews are full of accounts of drivers who abused their positions of authority, but Frank Bell recalled a different story: "Ole overseer on some plantations wouldn't let families work together, cause dey ain't gonna work as fast as when dey all mixed up, but Marse John Fallon had a black foreman, what was my mother's brother, my uncle. Moses Bell was his name, and he always looked out for his kinfolk."[14]

The plantation where Frank Bell grew up, however, seems to have been the exception to the rule in Fairfax County—at least among large plantations—as he himself admitted in his testimony. Most slaveholders believed that they could extract a maximum amount of labor from their work force by indeed keeping their slaves "all mixed up" in the fields. On small farms, however, especially toward the end of the antebellum period, family members were undoubtedly more likely to work together in the fields, if only because slaves were too few in number to divide into many squads. Indeed, on most small farms it was practically impossible not to work with at least a few family members.

Despite the testimonies of Steward and Bell, evidence suggests that enslaved people often worked with family members of the same sex— when they worked with family members at all—as many Fairfax County slaveholders preferred to employ a sexual division of labor in at least some of their agricultural operations. Plowing, for example, was almost exclusively performed by enslaved men. Males were also put in charge of jobs such as carting and clearing land. During the harvest, men were employed cutting and cradling the wheat. Women, on the other hand, were responsible for a number of tasks such as hoeing, raking and binding the harvested wheat, cleaning out stables, and even leveling ditches. Women and young girls were also more likely to be employed in the master's house as domestic servants if they could be missed in the fields. Increasingly, males and females worked separately under the system of mixed farming, although certain work sometimes called for slaves of both sexes to work in close proximity to each other, even if they were performing different tasks. During the wheat harvest the men would scythe and cradle, but the women followed close behind, raking and binding the cut grain. Plowmen were also closely followed by female family members in hoe squads. Yet with specific tasks often segregated by gender, despite spatial proximity between the sexes, it was more likely for enslaved family members of the same sex to assist or train one another in their work for the master—an arrangement that probably strengthened family ties along gender lines. Family structure of course played an important role. The extent to which fathers may have worked together with their sons obviously depended on local domestic arrangements—men whose wives and children lived on other holdings could not assist them in the fields. This would not have prevented them from working with their own brothers or extended male kin on the home farm, however.[15]

Flexible Parents and Spouses: Georgetown District, South Carolina

While enslaved field hands in Fairfax County were forced to labor day after day from morning till night, slaves on the rice plantations of Georgetown District enjoyed at least the opportunity of appropriating a portion of their day to spend with their family members, away from the watchful eyes of their masters and overseers. Indeed, the greatest advantage of the task system from the slaves' perspective was that it allowed industrious field hands to maximize their free time—time they could spend not as enslaved laborers for their masters, but as heads of their own households. But even during working hours enslaved people living along the rivers of Georgetown District were often afforded opportunities to maximize family contact, both in the fields and in the quarters. In all, task work and the nature of regional agriculture provided slave families with a unique degree of flexibility.

Rice planters in the lowcountry tended to treat their pregnant and newborn human property with more consideration than Fairfax County slaveholders, albeit more out of interest for the potential addition to their labor force than out of sympathy for the mothers' needs. The stagnant waters, oppressive heat and humidity, and rampant diseases of the lowcountry rice swamps posed serious health risks to pregnant women, unlike the wheat and corn fields of northern Virginia. As a precaution against disease and possible miscarriage, slaveholders in Georgetown District were especially keen to protect childbearing women from the unhealthiest features of rice cultivation. Women were usually assigned lighter work (half- or three-quarter-tasks) throughout their pregnancy and afforded an average lying-in period of about one month to recuperate after giving birth. Robert F. W. Allston, for example, preferred to give pregnant field hands light hoeing tasks in the upland provision crops. Plowden Weston encouraged pregnant women "to do *some* work up to the time of their confinement, if it is only walking into the field and staying there." But, he added, "if they are sick, they are to go to the hospital, and stay there until it is pretty certain their time is near." Once a child was born, Weston allowed new mothers a lying-in time of four weeks or longer, as necessary, followed by two weeks of light tasks on the upland. James Sparkman also took special precautions to ensure the good health of pregnant women and newborn infants on his plantation: "Allowance is invariably made for the women as soon as they report themselves *pregnant*, they being appointed to light work as will insure a proper consideration for the offspring. No woman is called out to work after her confinement, until the lapse of 30 days, and for the first fortnight thereafter her duties are selected on the upland, or in the cultivation of provision crops, and she is not sent with the gang on the low damp tide lands."[16]

Even with such precautions, however, mortality rates among both childbearing women and children under fourteen years of age were appalling in the stifling rice swamps. William Dusinberre has calculated that chronic epidemics of malaria in the lowcountry resulted in a large number of miscarriages, while relatively high mortality rates among female rice workers further reduced fertility rates to well below the southern average (6.6 children born to every slave woman in the lowcountry as compared to 7.5 in the South as a whole). Enslaved children in Georgetown District also fared significantly worse than their southern counterparts. Based on census information, Dusinberre calculated that the child mortality rate (of children aged 0–14) in Georgetown District was 66 percent in 1860, compared to a southern average of 46 percent. The plantation book of William Lowndes suggests that infant mortality rates were just as high at the beginning of the century. Of the fifteen slave children born on Lowndes's plantation in 1810, ten (67 percent) died within the first year.[17]

At the end of their lying-in periods, enslaved mothers had little choice but to return to work and share their child-rearing duties with others, just as they did in Fairfax County. However, shared child rearing took a distinct form on the rice plantations of Georgetown District: it tended to be less time-consuming and better organized in the lowcountry than in northern Virginia. To be sure, enslaved mothers and fathers on the rice plantations were forced to spend a significant amount of their time away from their young children, but in Georgetown District this amounted to an average of eight to ten hours per day, compared to sunup to sundown in Fairfax County. On Plowden Weston's plantation, slave women having "six children alive at any one time" were even allowed "all Saturday to themselves." Moreover, the unusually large numbers of slaves resident on most rice plantations meant that on virtually all estates, elderly women who were considered too old to perform field labor were available to look after young children while their parents were in the fields. The census returns for 1840, for example, indicate that in that year 89 percent of enslaved children under the age of ten lived on slaveholdings where enslaved women between the ages of fifty-five and one hundred were also resident. Most of the remaining 11 percent did not live on rice plantations at all but were rather the children of domestic servants in the village centers.[18]

On virtually all lowcountry rice plantations "young girls and old women minded the babies." In a few cases, even young adult women were put in charge of child care. On True Blue plantation during the Civil War, for example, a thirty-two-year-old bondswoman named Jane was appointed children's nurse along with three older women between the ages of fifty-five and sixty. Few enslaved children of field hands were therefore ever left unsupervised or under the care of domestic servants and white mistresses during the day, as they often were in Fairfax County. Indeed, most plantations contained a special nursery in the center of the slave village, where children were dropped off early in the morning and collected in the afternoon when their parents had finished their tasks. Captain Basil Hall observed on one such plantation: "It appears that when the negroes go to the field in the morning, it is the custom to leave such children behind, as are too young for work. Accordingly, we found a sober old matron in charge of three dozen shining urchins, collected together in a house near the centre of the village." Henry Brown, a former bondsman, recalled that on his plantation, "the babies were taken to the Negro house and the old women and young colored girls who were big enough to lift them took care of them." And J. Motte Alston, master of Woodbourne plantation, claimed that during slavery "there were houses where the children were cared for during the day when their mothers were at work." Nurses not only supervised small children but also taught them valuable skills. Elizabeth Allston Pringle of Chicora Wood plantation recounted how the elderly nurse "trained the children big enough to learn, teaching them to run up a seam and hem, in the way of sewing, and then stockings, and then to spin."[19]

The women and girls who worked in the nursery during the day were almost always the grandmothers, aunts, cousins, or sisters of at least some of the children under their care, which allowed them the opportunity to fulfill an important role in the upbringing of their own immediate or extended family members. Margaret Bryant, a former bondswoman, was quick to comment during an interview that the nurse on her plantation was her own aunt: "My pa sister, Ritta One, had that job. Nuss (nurse) the chillun. Chillun house. One woman nuss all the chillun while they ma in the field. All size chillun." Ellen Godfrey remembered taking care of her cousin in the nursery as a young girl: "Gabe Knox? I nurse Gabe. I nurse 'em. He pappy my cousin." Jane, the above-mentioned children's nurse on True Blue plantation, was fortunate enough to have her four children—ranging in age from two to twelve—under her care all day while her husband Daniel, a field hand, worked at his tasks in the rice swamps. To those children under their supervision who were not blood related, nurses undoubtedly served as fictive kin.[20]

Evidence suggests that the flexibility of task work afforded new mothers in Georgetown District several opportunities during the day to return to the nursery for breastfeeding as well, unlike in Fairfax County. One visitor noted that at the nursery on the rice plantation he visited in 1836, the "tender age [of the dry nurse] required the frequent visits of the mother." Olmsted likewise remarked that in the Georgia lowcountry new mothers "make a visit to them [their children] once or twice during the day, to nurse them, and receive them to take to their cabins, or where they like, when they have finished their tasks—generally in the middle of the afternoon." On some plantations, however, arrangements were made for the babies to be brought to their mothers in the fields to be nursed. Henry Brown told interviewers that "at one o'clock the babies were taken to the field to be nursed, then they were brought back to the Negro house until their mothers finished their work, then they would come for them." Elizabeth Allston Pringle likewise recalled that at Chicora Wood, "the nursing babies . . . were carried to their mothers at regular intervals to be nursed." On Plowden Weston's plantations, slave mothers were allowed to begin their tasks half an hour later than the rest in order to nurse their babies in the early morning.[21]

Well supervised back in the slave village, there was little reason for masters or overseers to keep enslaved children past infancy out of mischief by employing them in the fields, as they often did in northern Virginia. Health risks played a role as well. The rice swamps were dangerous for young children, and most aspects of rice cultivation were too physically demanding for them to perform anyway. When the need arose, however, overseers and masters did not hesitate to occasionally employ their help in performing light, yet important, chores. Such chores not only contributed to the daily functions of the plantation, but they also accustomed future field hands to the adult world of work that would one day dominate their lives. Olmsted observed, for instance, that some children were sent to the fields during the day and "charged with some kind of light duty, such as frightening birds." He also noted that in each field "a boy or girl was also attached, whose business it was to bring water for them [the field hands] to drink." Older children were sometimes assigned chores in the slave village, the most common of which for girls was helping the elderly women to supervise the smaller children in the nursery. In general, however, enslaved boys and girls on the rice plantations were not expected to work regularly until they were between the ages of twelve and fourteen, when they entered the work force as quarter- or half-task hands. On True Blue plantation the youngest field hands were even sixteen years old.[22]

The flexibility that accompanied task work also allowed able-bodied members of slave families to work together in the rice fields. Historian Leslie Schwalm found that "while each field hand was assigned her or his own task, the completion of a day's task was often a cooperative effort among slaves in the field." The division of labor employed on the rice plantations especially allowed for family members of the same sex to assist one another in completing their daily tasks, as gender segregation, in the words of historian Emily West, "restricted opportunities for male-female bonding while toiling for the master." James Sparkman, for example, declared that "in the preparation of the Rice Lands, as ditching, embanking, etc. the *men* alone are engaged with the spade." Jesse Belflowers, the overseer at Chicora Wood, reported to his employer on 28 June 1862 that the "Men [are] working in the Pine land, the women picking rice." Such divisions between male and female laborers was common throughout the lowcountry, not just in Georgetown. Olmsted observed on a rice plantation in the Georgia lowcountry some "twenty or thirty women and girls . . . engaged in raking together," while "in the next field, twenty men, or boys, were ploughing." In practice, such arrangements afforded slaves opportunities to assist family members of the same sex, which they often seized. Hagar Brown, a former bondswoman from Georgetown District, recalled her mother often coming to the assistance of her aunt in the rice fields in order to protect her from corporal punishments. "Beat ma's sister. Her sister sickly. Never could clear task—like he [the overseer] want. My ma have to work herself to death to help Henritte so sickly. Clear task to keep from beat[ing]. . . . Ma had to strain to fetch sister up with her task. Dere in rice field."[23]

Such assistance was not strictly limited by gender boundaries, however, which set the lowcountry apart from northern Virginia and many other parts of the South. In the afternoons, men regularly came from completing their own tasks to help their wives and other family members. One traveler reported that upon completion of their tasks, men "were allowed to come to the aid of a wife or a friend, who had been less fortunate in bringing up the labors of the day." A local rice planter recalled seeing "with much pleasure the husband assisting the wife after he was finished with his own task, and sometimes I have seen several members of a family in like manner." James Sparkman informed a prospective planter that "it is customary (*and never objected to*) for the more active and industrious hands to assist those who are slower and more tardy in finishing their daily task." By coming to the aid of family members, slaves not only protected them from corporal punishment, as suggested by Hagar Brown, but they also maximized the amount of family time they could spend back in the slave village.[24]

Heavy Burdens: St. James Parish, Louisiana

The forced nature of cane cultivation had decidedly negative consequences for slave families living in St. James Parish, Louisiana. The intensity with which gang labor was applied to field work on the sugar plantations, which some likened to military drill, demanded especially long and physically exhausting work days of the slaves. Toiling as they did at a furious pace from sunup to sundown, six days a week (and sometimes seven), slave families had little time indeed to devote to child rearing or quality time with family members. During the grinding season, when round-the-clock shifts were expected of all able-bodied slaves for two to three months on end, free time was for all intents and purposes nonexistent. Yet despite their long work days, enslaved people's daily lives in the sugar country differed in many ways from their counterparts in Fairfax County. Indeed, the relationship between work and family life in St. James Parish combined characteristics found in both northern Virginia and lowcountry South Carolina.

The experiences of pregnant women and newborn infants on the sugar plantations provide a case in point. As on the rice plantations, most planters in the sugar country took special precautions to ensure the health of their pregnant bondswomen, who were singled out to perform lighter work than the other field hands. However, when lighter workloads were weighed against the pressing demands of sugar cultivation, the latter frequently won out over the needs of expecting mothers. For example, pregnant women in St. James Parish were often assigned lighter chores not immediately after they reported themselves pregnant, as in Georgetown District, but rather when they were well into their final trimester, which more resembled the experiences of women in Fairfax County. Indeed, evidence suggests that some plantation managers were not attentive at all to the reproductive needs of enslaved women. Edward De Buiew, a former bondsman, told interviewers of the Louisiana Writers' Project in the 1930s that his "ma died 'bout three hours after I was born. Pa always said they made my ma work too hard. I was born in the fields. He said ma was hoein.' She told de old driver she was sick; he told her to just hoe right on. Soon, I was born, and my ma died a few minutes after dey brung her to the house." At Houmas plantation in 1853 a woman named Isabell was confined only one day before she gave birth. Historian Richard Follett has argued that many sugar planters "favored labor over leave" and kept pregnant women in the fields until shortly before they gave birth, especially on newly established plantations with limited work forces.[25]

Many of the wealthier and more progressive sugar planters, however, as many of those living in St. James Parish undoubtedly were, were keen to protect enslaved women in advanced stages of pregnancy from overwork, because overexertion in the stifling and unhealthy cane fields could result in miscarriage and thus endanger a valuable potential addition to their labor force. John Burnside, one of the wealthiest sugar planters and largest slaveholder of Louisiana, took special care in preventing miscarriage. His overseer boasted to British reporter William Howard Russell that "there is not a plantation in the State . . . can show you such a lot of young niggers. The way to get them right is not to work the mothers too hard *when they are near their time*; to give them plenty to eat, and not to send them to the fields too soon." As a result of this policy, Burnside's slave population allegedly increased naturally by 5 percent each year. Other planters in the sugar country took similar precautions. Historian V. Alton Moody found that on most large sugar plantations, expecting mothers were eventually "assigned to only light duties."[26]

As in the lowcountry, slaveholders in southern Louisiana often afforded new mothers an average lying-in time of about one month as a health precaution. The daughter of one local sugar planter recalled that on her plantation "no woman went to work until her child was a month old." The amount of time given to new mothers to recuperate was, however, flexible and subject to the opinion of the master. Maternal leave could be shortened if the health of both mother and child appeared secure, but on Houmas Plantation in neighboring Ascension Parish, new mothers were occasionally allowed more than a month if necessary. One enslaved woman named Louisa bore a child on 7 January 1853, and did not return to the fields until 1 March, most likely because of complications. Confinement periods were no holiday, however, as it was not uncommon for new mothers to be required to perform certain tasks even during their lying-in time. The sugar plantation that could afford idle slaves, even those temporarily incapacitated for field work, was rare indeed. Former bondswoman Ceceil George remembered that "when a woman has a baby—if she can't go to the field—when de baby is nine days old she has to sit and sew." A local sugar planter admitted to agriculturalist Solon Robinson in 1849 that all of the clothing on his sugar plantation was "spun and woven by . . . mothers, just before and after giving birth to children." When they did finally return to the fields, new mothers were temporarily assigned to a so-called "sucklers gang" and eased back into the regimentation of cane cultivation with relatively light workloads, similar to the experiences of women in Georgetown District.[27]

Unlike their counterparts in the rice country, enslaved women working according to the gang labor system lacked the flexibility to be able to return to the quarters to nurse their newborn infants, but planters and overseers on the sugar plantations usually granted new mothers explicit permission to do so during working hours, as is evident from a number of travelers' accounts. William Howard Russell noted while visiting a plantation in St. James Parish that new mothers were "permitted to go home, at appointed periods in the day to give the infants the breast." Elsewhere in the sugar country, Francis and Theresa Pulszky observed slave "women suckling their babies" during a brief break in the day. And Olmsted likewise reported meeting "half a dozen women, who were going . . . to suckle their children—the overseer's bell having been just rung (at eleven o'clock), to call them in from work for that purpose." On this particular plantation new mothers were allowed two hours to spend with their babies, but this appears to have been uncommon and probably amounted to only an extended midday meal break.[28]

Indeed, on most sugar plantations the amount of time allotted to new mothers to nurse their babies was modest at best, and slave mothers did sometimes risk punishment to push the limit. Former slave Charlotte Brooks told an interviewer in the late nineteenth century that her master "did not allow me much time to stay with my baby when I did go to nurse it. Sometimes I would overstay my time with my baby; then I would have to run all the way back to the field." The period for breastfeeding newborns did not last long, either. Infants in St. James were weaned relatively early and put on a diet of "pot liquor and mush and molasses," according to the daughter of one planter. As Richard Follett has argued, such practices may have simply been the result of sugar planters putting the pressing demands of cane cultivation above the needs of newborn infants, but planters also may have intentionally sought to trigger "an early return to normal luteal function and ovulation"—in other words, to induce shorter birth intervals and thereby maximize the reproductive potential of their enslaved labor force.[29]

Despite their need for labor over leave, sugar planters had good reason to want to provide basic care to new mothers and their infants, as most local plantations struggled with a severe shortage of children and an insatiable demand for labor, making each new birth a valuable financial asset. Indeed, in 1850 slave deaths exceeded births by 28 percent in St. James Parish. Historian Michael Tadman has argued that factors such as high infant mortality due to disease and excessive labor among pregnant women, combined with a severe sexual imbalance (as shown in chapter 5), contributed to the unusually low number of children on sugar plantations. Ira Berlin calculated that the fertility rate of enslaved women in St. James in 1850 was a meager .54 (or .54 children under the age of five for every enslaved woman between the ages of fifteen and forty). This was 40 percent lower than the fertility rate in the cotton regions. The scant number of children listed in numerous local slave inventories seems to corroborate these findings. Only twenty-nine children of all ages appeared in the 1860 slave inventory of W. P. Welham's plantation in St. James Parish, for example, despite some thirty-two resident slave women of childbearing age—and three of these children were classified as orphans. Samuel Fagot's plantation contained only thirty-four children under ten years of age in 1855, despite having forty-four resident women of childbearing age. High child mortality rates and below-average fertility did not go unnoticed by visitors to the region. Philo Tower reported that the work of women was "so hard, that very few children are raised on the sugar . . . plantations; and if they are alive at birth, they grow up feeble and puny, and . . . very few become men and women."[30]

Progressive sugar planters were thus keen to prevent unnecessary infant deaths on their plantations, but in this they were not always successful. On the frequent occasion that an infant did die, slaveholders often unleashed their frustration by accusing slave mothers of improper care. Valcour Aime recorded that the death of one baby named Washington in 1839 was due to "bad nursing by his mother." The death of another six-month-old was attributed to its "mother [being] too dumb to raise it." Other new mothers were accused of dropping their infants and smothering them in their sleep. Olmsted noticed a similar attitude of anxiety and distrust toward slave mothers during his visit to a sugar plantation. Meeting a group of new mothers as they returned home from the fields to nurse their babies, his planter-host asked the women, "The children all well?" When the answer came that all were well except for the child of one Sukey, the planter became irritated. "Sukey's? What, isn't that well yet? . . . But it's getting well, is it not?" If the mother was not the culprit, then diseases and complications surely were. Aime recorded numerous infant deaths due to teething, fevers, spasms, pleurisy, cholera, and whooping cough.[31]

At the end of their lying-in period, enslaved women had little choice but to leave their newborn infants behind as they were summoned back to the cane fields. As in Fairfax County and Georgetown District, thus, child rearing was primarily a shared responsibility on the sugar plantations of St. James Parish. Indeed, long and arduous workdays even conspired to make enslaved parents there more dependent on others to supervise and raise their children than perhaps anywhere else in the slave South. As in Georgetown District, the relatively large size of many slaveholdings in St. James Parish often provided plantations with at least a handful of resident elderly women who could look after the children while their parents were in the fields. However, the percentage of enslaved children who lived on such estates did not approach that reached on the rice plantations in Georgetown District. The census returns for 1840 show that in St. James Parish only about 36.5 percent of slave children lived on holdings that contained elderly women who could have cared for them during working hours. When the figures are adjusted to exclude holdings with fewer than five children (which were probably not farms or plantations but households with domestic servants), however, the percentage rises to 55 percent. This is compared to 89 percent in Georgetown District, but 34 percent in Fairfax County.[32]

Indeed, the slave population of St. James Parish was overall quite young, mainly because southern Louisiana was still a developing slave society. The mass importation of able-bodied slave laborers only began to take shape in the opening decades of the nineteenth century, and both new and established plantations continued to be regularly injected with young imports from the eastern states as late as the Civil War. Therefore, most of the enslaved men and women of the "migration generation" who were brought to the parish to toil on the sugar plantations were not yet past their prime in 1840. High mortality rates, as suggested by Tadman and others, exacerbated the situation. Census returns indicate that in the 1850s, some 27 percent of slaves between the ages of forty and forty-nine died before their fiftieth birthday. Fully half of slaves aged fifty to fifty-nine died before their sixtieth birthday. Valcour Aime recorded the deaths of numerous adult slaves on his plantation, most of whom fell victim to cholera, fevers, dropsy, consumption, and aneurisms. Pam Drake expressed revulsion during her tour through the sugar country at the admission "by slaveholders themselves, that of a hundred young men, taken from Kentucky and Virginia, and put on plantations in Louisiana . . . seven years will be the average of their lives." This was surely an exaggeration, but in any case a number of relatively large sugar plantations in St. James Parish contained no elderly slaves. In 1840, George Mather's plantation contained 115 slaves, not one of whom was older than fifty-four. R. Bell's estate counted 136 slaves—again, all under fifty-five. For purposes of child care, this was clearly not an advantageous situation for enslaved parents, although in some cases women younger than fifty-five were put in charge of caring for the children. On W. P. Welham's plantation, a forty-six-year-old woman named Fanny, who had a "sore leg" and was thus incapacitated for field work, was made a children's nurse.[33]

Eager to provide adequate child-care facilities to protect valuable future additions to their enslaved labor forces, most of the larger sugar planters in St. James Parish built nurseries on their estates where slave children were dropped off every morning at dawn and picked up again at night when their mothers and fathers returned from the fields, as was the case in Georgetown District. These nurseries were commented upon by visitors, residents, and former slaves alike. A guest at one local plantation in 1852 reported that the children on the estate were kept in "a nursery fifty feet square" and looked after by "four nurses" during the day. Eliza Ripley, the daughter of a local sugar planter, recalled in her memoirs that on her plantation there was "a day nursery for the babies, under the charge of a granny . . . there the babies were deposited in cribs all day while their mothers were at work in the fields." William Howard Russell found that in St. James Parish, "those who are mothers leave their children in charge of certain old women, unfit for anything else." And former bondswomen Francis Doby and Rebecca Fletcher also told interviewers that there were nurseries on their plantations. Doby claimed that "de old women—too old to work and too old to make de babies—dey stay and mind de young children so dat de ma can all work in de fields." Fletcher recalled that during the day "the children were left behind. An old woman had the care of 'em, and it was in a big kitchen where she cooked and fed 'em."[34]

As in Georgetown District, these nurses were usually related to at least some of the children under their care, affording some elderly women the opportunity to play an important role in socializing their young kin and passing on valuable life skills. One ex-slave named Melinda remembered that she remained in the nursery "with my grandma from sunrise to sunset. So it is that I was very fond of her and learned many useful things, for she knew the value of herbs and how to prepare remedies for almost every evil." The elderly thus played a vital role in child rearing on such sugar plantations, both as caretakers and as teachers, as they passed on their knowledge to the young children under their supervision. In some cases the children's nurses were even assisted by elderly men. Francis Doby claimed that on her plantation, a man who was too old to work in the fields and of whom the children were especially fond would frequently visit the nursery and amuse them by teaching them to drum and dance. To those children who were not blood related, elderly caretakers often served as fictive kin. Doby, who was sold away from her real grandmother as a small child, called the nurse who looked after her on her new plantation "Grandma."[35]

On slaveholdings that did not contain any elderly women who could look after the children while their parents were in the fields, alternatives had to be found, not unlike the experiences of enslaved parents living on the small farms of northern Virginia. In St. James Parish women younger than fifty-five who were incapacitated for field work were sometimes made children's nurses, as was the case on Welham's plantation. Charlotte Brooks, however, claimed that temporary caretakers were sometimes drawn from the slave labor force: "old mistress would have some one to mind [the children] till they got so they could walk, but after that they would have to paddle for themselves." The testimony of Daffney Johnson, a former bondswoman, reveals that young girls were also often put in charge of supervising the younger children. Her first occupation as an adolescent was "to mind the children," she told interviewers. Occasionally, it appears that small children were loosely supervised by the slave cooks who prepared the midday meals for the field hands. Francis and Theresa Pulszky visited the kitchen of one slave village while the cooks were busily preparing dishes of pork with rice and vegetables for the hands. There they observed a "host of little black imps hovered around the kitchen, eating their hot corn bread. They were very dirty, and appeared to delight as much in mud as any gipsy."[36]

Even if many enslaved children in St. James Parish appear to have been well supervised during the day, their parents were deprived of spending much time with them by extremely long and physically exhausting workdays in the cane fields. During the grinding season, when eighteen-hour days were expected of able-bodied slaves, enslaved parents had virtually no time at all to spend with their young children. On the sugar plantations of St. James Parish, thus, shared child rearing took on its most extreme form in the slave South. Visitor E. S. Abdy pitied the enslaved mothers on the sugar plantations which he visited, as they were "prohibited from seeing [their infants] till their return at night." The testimonies of former slaves confirm Abdy's observations. Francis Doby's experience was that the nurses "feed [the children] and all so when dey ma come back, all dey got to do is to push dem in de bed, all of dem in de same bed." Elizabeth Ross Hite told interviewers that the "mothers did not have time to take care [of their children]" on her plantation. Hunton Love recalled that "when [the] day's work was over, we was too tired to do anything but go to sleep." The lack of time and energy that parents had to contribute to child rearing was the source of much frustration in the sugar country and may have been the reason why after freedom came, according to Rebecca Fletcher, the women all "took the babies along to the fields," rather than leave them behind for anyone else to care for.[37]

For most children past infancy, weekdays were spent fishing and playing games while their parents were in the fields. Children living on the sugar plantations of St. James Parish were rarely required to perform regular field work until a relatively late age, much like children in the rice country. The reasons were the same as in Georgetown District: cane cultivation was too physically demanding for children, and the appalling mortality rates in the region convinced most planters to leave them in the quarters until they were strong enough to wield a hoe. Elizabeth Ross Hite said that her master "didn't want no children to work. He used to say all de time, 'Don't let dem little darkies work. It might hurt dem, and dere is enough of dem older darkies on dis farm for work.'" Plantation records reveal that on W. P. Welham's plantation in 1860 the youngest field hand was fourteen years old. During his visit to the parish, William Howard Russell was "glad to see the boys and girls of nine, ten, and eleven years of age [on the plantation] exempted from the cruel fate which befalls poor children of their age in the mining and manufacturing districts of England." On another plantation he observed that of the children left behind in the quarters each day, "some [were] twelve and some even fourteen years of age."[38]

This did not mean that children were completely exempt from labor on the sugar plantations, however. Light tasks were expected of them during especially labor-intensive periods such as the grinding season, as was the case in northern Virginia and lowcountry South Carolina. Steven Duncan, a former slave, said that his first taste of plantation work came when he was six years old, when he was made to carry meals to the mechanics in the sugar house during the grinding season. Young boys were also sometimes stationed at the mill to keep it free of tangled stalks. Ex-slave Ceceil George recalled that during the cane-cutting season on her plantation, children were required to help carry the stalks to the sugar house: "If you could only carry two or three sugar cane [stalks], you worked." Other planters began to train their slaves between the ages of eight and ten, as they gradually assigned enslaved children an increasing number of responsibilities in the fields. Fred Brown told interviewers, "When I'se about 8 years old, or sich, dey starts me to helpin' in de yard and as I grows older I helps in de fields." Edward De Buiew remembered "the first work I ever done is to carry water to de fields." He also recalled sitting "out in de corn patches and mind de crows out of de corn." In practice, such tasks in the fields and sugar house may have provided children with opportunities to briefly spend time or even work with family members during the day, although specific evidence is lacking.[39]

As in Fairfax County, some slave children were singled out to perform domestic tasks in the house of the white family until they were old enough to be put to useful labor in the cane fields. Many assisted the domestic servants with the cooking or washing. Joseph Holt Ingraham observed on one plantation a group of children "conveying water from a spring to the washhouse, in vessels adroitly balanced upon their heads." Others were made personal servants to the white family and their guests. Thomas Hamilton, a guest on one sugar plantation, was waited on by a "very nice-looking black boy, who, after setting down my candle and adjusting the pillows of the bed, still remained standing right opposite to me when I began to undress." Daffney Johnson, enslaved on a sugar plantation, told interviewers that she "worked around de house a long time fore de boss put me in de sugar cane fields." And Melinda, likewise a former bondswoman, recalled that when she was eight years old, "dey took me in de house wid de white folks. [They] made me some pretty gingham dresses and told me I had to tend to Miss Dee-Dee, de youngest daughter of Madame Charles de Villere."[40]

Once slaves were put to regular gang work in the fields, usually by the age of fourteen, they were often confronted with a sexual division of labor, much like their counterparts in Virginia and South Carolina. Gang labor on the sugar plantations closely resembled assembly line production, and its application as such favored a strict division of labor according to ability, which most planters interpreted to mean sex. Therefore, the extent to which slaves in St. James Parish may or may not have been permitted to work in family groups must be viewed from the perspective of periodic gender segregation in the workplace.

Plow gangs in the cane fields, as in Georgetown District and Fairfax County, consisted almost exclusively of men. William Howard Russell noticed "one gang of men, with twenty mules and ploughs . . . engaged in running through the furrows between the canes." Hoe gangs often consisted of both men and women. Valcour Aime recorded on 3 June 1850: "Employing the whole gang of laborers at the hoe." However, even these gangs were occasionally segregated by sex. On Governor Roman's plantation, a "gang consisting of forty men . . . were hoeing out the grass in the Indian corn. [Another] gang, of thirty-six women, were engaged in hoeing out cane." During the harvest season men often performed the cane cutting, while women (and children) gathered the stalks and transported them to the sugar house. For the myriad of other tasks performed on the sugar plantations, slaves were likewise divided into sex-segregated gangs. Octave Colomb recorded in his plantation journal in February of 1850: "Men in ditches. Women clearing new ground." Aime repeatedly sent female gangs to clean and dig canals, work on the levee, cut weeds and hay, gather corn, and even repair roads. Male gangs chopped and hauled wood, dug ditches, and cut hay. Moreover, men dominated virtually all of the skilled positions on the sugar plantations, from blacksmithing to sugar boiling. Evidence does not indicate whether slaves were permitted to work alongside their immediate family members during their time for the master, but if they did, they usually worked with family members of the same sex.[41]

Conclusion

The nature of work in the slave South limited the amount of time that enslaved people had to spend with their family members during working hours. However, the specific demands of local cash crop cultivation resulted in different experiences for enslaved families across space, resulting in a variety of boundaries and opportunities for slaves in different slave societies.

While child rearing was everywhere to some extent a shared responsibility within slave communities, its most extreme forms appeared in northern Virginia and southern Louisiana, where enslaved parents were kept in the fields from sunup to sundown and deprived of valuable bonding time with their children. Only on the rice plantations of Georgetown District, where task work prevailed, were enslaved mothers afforded the opportunity to limit their dependence on substitute child rearers to the morning hours and early afternoons. They were also granted more flexibility to return to their small children during the day to nurse them, and the size of slaveholdings in the region afforded them the opportunity to place their children under the care of elderly women whose sole responsibility was to take care of the youngsters. This was not always the case in Fairfax County or St. James Parish, although in the latter county nurseries were often available and new mothers were usually granted brief visits to their infant children to nurse them.

Out in the fields, enslaved families in each of the three regions faced a division of labor that segregated men and women in the workplace. This strengthened bonds within families along gender lines, as family members of the same sex tended to come to each other's aid while working for the master. In Fairfax County, male and female family members often worked in close proximity, even if they performed different tasks, and as the antebellum period wore on many labor forces became too small to prevent at least some family members from working together. Again, on the rice plantations enslaved men and women were afforded the best opportunities to systematically cross gender lines when helping their family members to finish their tasks—opportunities they readily seized. As the following chapter shows, the unusual flexibility that characterized enslaved people's work in the Carolina lowcountry extended into the development of family economies in their free time as well.

4

Family-Based Internal Economies

Enslaved people's extreme dependence on their masters for shelter, clothing, and food—all of which were provided in standardized rations, regardless of how much individual slaves actually produced—theoretically placed them outside the realm of the daily struggle that characterizes free labor. In their own time, however, slaves did utilize a number of strategies to supplement their rations, acquire material luxuries, and make their lives more comfortable. Engaging in what scholars have often termed "independent production," slaves often developed limited, in some cases even extensive, family-based economies during their free time. Activities such as cultivating family garden plots, hunting, fishing, or voluntarily hiring out their labor in exchange for cash may not have constituted a struggle for physical survival, but they did provide slave families with opportunities to work together as economic units and enjoy the fruits of their own labor, thereby strengthening family ties and enriching the family diet. Some historians have argued that such economies also allowed slaves to carve for themselves a "niche of autonomy," which nourished their desire for independence and even stimulated "proto-peasant" behavior—as slaves, like peasants, made their own decisions about what and how much to produce, and how to dispose of their marketable products.[1]

Proto-peasant behavior may have been widespread in the Caribbean, where slave economies were highly developed, but in the American South enslaved people's formal work (their work for the master) and informal work (their work for themselves) do not appear to have been two separate work systems. The term "internal production" is perhaps more accurate than "independent production" in describing enslaved people's private economic activities, since these economies clearly did not develop *outside* of the institution of slavery, or *despite* it, but rather *within* its very walls and *as part and parcel* of its structure. As this chapter shows, family-based internal economies were inextricably interwoven with the broader demands of slave-based agriculture and especially with such factors as work patterns and master-sponsored labor incentives, giving slaveholders the ultimate say—both directly and indirectly—in slave families' opportunities to acquire small luxuries.[2]

How did the demands for cultivating local cash crops and the nature of regional agriculture affect the development of family-based internal economies among the various slave populations? And what does this tell us about the "autonomy" of such economies? The answers to these questions are further examined in this chapter.

Arrested Development: Fairfax County, Virginia

When Edward Dicey, a British journalist, passed through Fairfax County soon after the outbreak of the Civil War, he found the enslaved population there living under what appeared to be abject poverty. In his journal he jotted down that local bondspeople were "miserably clothed, footsore and weary." Encountering a group of runaways, he and his men offered them some white bread, upon which one of the fugitives remarked, "Massa never gave us food like that." Dicey later recalled, "Anything more helpless or wretched . . . I never saw." Material conditions for slaves in northern Virginia may have left much to be desired in the first place, but the situation was doubtless exacerbated by an underdevelopment of family-based internal economies. Working from sunup to sundown, men and women in bondage had little time off during which they could try to supplement their rations or acquire small luxuries by cultivating family garden plots, hunting, fishing, or voluntarily hiring out their labor in exchange for cash. Not only were enslaved people's standard work patterns detrimental to families' material conditions in Fairfax County, however, but the general economic decline in local agriculture dealt them an additional blow by saddling the diminishing slave population with even heavier workloads, limiting opportunities to develop highly skilled trades, and convincing slaveowners to curb internal production, often at the cost of traditional privileges to work for themselves in their free time.[3]

Opportunities for local slaves to voluntarily hire out their labor in exchange for cash were severely circumscribed. A majority of the skilled artisans in the region—those in the best position to hire out their services in their free time—lived not on farms but in towns such as Alexandria, Fairfax, or Centreville. In the countryside, where most enslaved people lived, only some of the larger slaveholdings contained craftsmen who were able to occasionally profit from their trades, especially in the early decades of the nineteenth century. General Thomson Mason, for example, owned about sixty slaves in 1810, including "his weaver, his blacksmith, his carpenter, [and] his shoemaker." These artisans were probably able to occasionally profit from their skills by producing for outside clients. As the antebellum period progressed and the number of large slaveholdings declined, however, the number of skilled slaves living in the countryside dwindled. Those who remained were often forcibly hired out by the year in urban areas, where they could make their masters more money.[4]

A vast majority of slaves looking to perform paid overwork were ordinary farm hands whose best options consisted of odd jobs and day labor. In some instances, local slaves were called upon by their masters to perform extra work for cash after sundown. Former bondsman Austin Steward claimed that on his plantation enslaved men "were occasionally permitted to earn a little money after their day's toil was done." Others—especially men—sought to voluntarily hire out their services to local farmers during their limited free time; in other words, on Sundays, pending their master's permission. Ever motivated to improve their own and their families' material conditions, some men certainly seized the opportunity whenever it presented itself. One northern Virginia slave named Silas Jackson, for example, recalled that "when we could get work, or work on somebody else's place, we got a pass from the overseer to go off the plantation, but to be back by nine o'clock. . . . Sometimes we could earn as much as fifty cents a day, which we used to buy cakes, candles and clothes."[5]

Such work was hardly regular, however, and often limited to slack periods in the agricultural cycle. Former slave George Jackson claimed that during most of the year the men on his plantation "would sit around" on Sundays, apparently resting and socializing, but not working. Solon Robinson, who visited Mount Vernon on one particular Sunday afternoon in June of 1849, observed a number of "lazy negroes listlessly hanging about," obviously none of whom were performing paid work.[6]

Even those who did earn extra money accumulated relatively little, as time was limited and opportunities were few and far between. Contrary to the claims of some local historians that "many slaves bought their liberty with money they earned while hiring their own time," little evidence suggests that many resident field hands in Fairfax earned enough money to purchase their own freedom. Indeed, the antebellum Fairfax County Deed Books explicitly mention only two cases of slaves who managed to save enough money to buy themselves out of bondage. In 1824 one slave woman named Abbey bought her freedom for the token sum of $40, which she earned by her "services" to her master. In reality she was probably elderly or otherwise unfit for productive labor, as the deed also mentions that Abbey must "account and do for herself" from that day forward, making it sound as if her master simply wanted to be rid of her. In 1841, local bondsman Amos Nickens bought his freedom from his master, one John McDonald, for $200. Virtually all other slaves who obtained their freedom legally were manumitted by the deeds or wills of their masters, or bought and freed by free black family members. Although in Virginia manumission depended on a number of complicated legal factors, few local enslaved people would have had access to the sum that Nickens paid for his freedom. Local slaves who did manage to earn extra cash were usually forcibly hired out by the year and permitted to keep a small portion of their earnings, most likely including Nickens himself.[7]

Local family structures may have further discouraged married slave men from hiring out their labor on Sundays. Although slave family structure is not the primary focus of this chapter, but rather of chapter 6, its significance to the nature of slave family economies deserves brief mention here. The dwindling size of agricultural units throughout the antebellum period in Fairfax County and elsewhere in northern Virginia resulted in a prevalence of cross-plantation families, by which men lived apart from their wives and children but usually visited them on their days off. Under such circumstances it is plausible that enslaved men preferred to visit their loved ones on the weekend rather than hire out their labor for cash. Former slave George Jackson recalled that on "Saturday night those [enslaved men] dat hed wives would go to see them," usually returning Monday morning for work in the fields. He did not mention that any performed paid overwork.[8]

Instead of hiring out their labor in exchange for cash on Sundays, enslaved men in northern Virginia chose alternative ways of supplementing their families' diet, methods that required less time and energy and did not cost them valuable family time. In the evenings, many enslaved men procured what meat they could from the woods and creeks, trapping rabbits, opossum, and other small game. One enslaved field hand living near Alexandria told a traveler that his master issued his slaves only cornmeal, and that "for fifteen months [I] did not put a morsel of any meat in my mouth, but the flesh of a possum or a racoon that I killed in the woods." More frequently, perhaps, enslaved men fished in the Potomac River or one of the many creeks in the county. One visitor to Fairfax County observed on the Potomac "one or two negroes stationed at the bottom of the precipice on a point of a reef, fish[ing] for *shad*, with nets attached as a sack at the end of a pole. I watched them for a half hour dipping and retrieving their nets but always without success." Fish, rabbits, and opossum could be consumed by the family or (illegally) sold for cash.[9]

Cross-plantation marriages, combined with long workdays, also adversely affected the division of labor within slave families' internal economies, especially as they related to family garden cultivation. With many husbands and fathers absent from slave family households during the week, domestic responsibilities such as maintaining family garden plots fell to women, who already had enough to do. Former slave Francis Henderson claimed that early in the morning and late at night, when they were not working for the master, enslaved women on his farm had time for little else besides cooking or tidying up the house, which was always "dirty and muddy." Henderson also recalled that slaves on his farm often had to share cooking fires with others, causing fatigued women to have to wait in line "before [they] could get to the fire to bake hoe cakes." Optimal cultivation of the family garden plot may not have been an enslaved woman's first priority in a region where most of her severely limited "free time" was tied up in more pressing duties, such as child rearing, washing, cleaning, sewing, and cooking. Historian Larry Hudson has found that slave families with few resident able-bodied members often failed to adequately work garden plots and accumulate property in the American South. Cross-plantation families in Fairfax County appear to have fit this pattern. Keeping poultry may have been an easy way of raising extra food for those with little time or energy, but sources do not indicate how widespread this practice was in Fairfax County or to what extent it was allowed.[10]

In fact, access to family garden plots at all seems to have been limited among enslaved people in northern Virginia. Visitors to the region who were quick to comment on the existence of slaves' garden plots in other parts of the antebellum South curiously omitted any mention of their existence in Fairfax County. A few antebellum visitors to nearby Alexandria mentioned the presence of African Americans at urban places of market, an indication that perhaps local enslaved people were permitted to sell their own produce on the weekends. Frederick Law Olmsted, for instance, remarked that "the majority of the people [at the Alexandria market] were negroes." Traveler Robert Sutcliffe likewise reported that Alexandria teemed with slaves from the country selling "fruit, vegetables, &c." However, such testimonies can be misleading. First, it is unclear whether the blacks mentioned by Olmsted were enslaved or free, as Alexandria had a sizable free black population in the antebellum period. Second, some of the country slaves at local markets, as mentioned by Sutcliffe, may not have been selling their own produce but that of their masters. With the advent of mixed farming in the region, some slaveholders tried to raise extra cash by occasionally sending one or two of their slaves to Alexandria to sell produce grown on the farm. This explanation seems more viable when one considers that these African Americans were observed selling their produce out of "rickety carts, drawn by . . . oxen and horses." It is doubtful that any but the most privileged slaves would have been granted permission to borrow carts, oxen, or horses from their masters in order to sell their own produce. Robert Sutcliffe spoke to a slaveholder living near Washington who regularly sent the driver of his farm to town with a "wagon laden with various kinds of produce, the sale of which he entrusts to this black man; also the care of bringing home the money; by which means it frequently happens that large sums of money pass through his hands." Olmsted heard a similar story from a local man: "Mr. F said that an old negro woman once came to [him] with a single large turkey, which she pressed him to buy. [He] ascertained that she had been several days coming, had travelled mainly on foot, and had brought the turkey and nothing else with her. 'Ole massa had to raise some money somehow, and he could not sell anything else, so he tole me to catch the big gobbler, and tote um down to Washington and see wot um would fetch.'"[11]

Little evidence suggests that the privilege of cultivating family garden plots was universal or even widespread among enslaved people in Fairfax County. Quite the contrary. George Jackson remembered that "de slaves on our plantation did not own der own garden. Dey ate vegetables out of de big garden." James Smith, enslaved in nearby Northern Neck, recalled that the only supplement to his family's rations came from "crabbing and fishing," but not gardening. One Fairfax County slave named Randolph, who fled to Canada, told interviewers upon arrival that he had worked for his master "without privileges." And a visitor to Mount Vernon around the turn of the nineteenth century observed next to one of the slave cabins "a very small garden planted with vegetables," and also "5 or 6 hens," but added that the slaves were strictly prohibited from keeping "ducks, geese, or pigs," and that Washington seemed to "treat his slaves far more humanely than do his fellow citizens of Virginia. Most of these gentlemen give their Blacks only bread, water and blows." It appears that in many cases, garden plots were offered only as rewards to particularly hard-working slaves. Austin Steward, enslaved near the border of Prince William and Fairfax at the turn of the nineteenth century, recalled in his autobiography that "some slaves were permitted to cultivate small gardens, and were thereby enabled to provide themselves with many trifling conveniences. But these gardens were only given to some of the more industrious."[12]

Fairfax County slaveholders' reluctance to offer all of their slaves family garden plots seems to contradict the belief among many planters in other parts of the South that such privileges produced healthier, more efficient, and less rebellious workers. Indeed, one scholar has calculated that around 60 percent of former slaves who were interviewed by the Federal Writers' Project mentioned having had access to family garden plots. In Fairfax County, however, a number of economic and social factors may have discouraged local slaveholders from employing the "work and garden" incentive. First, the nature and profitability of the local slave-based economy proved in some ways incompatible with gardening privileges. Second, social realities in the devolving slave society appear to have induced slaveholders to tighten their controls against slaves' internal production.[13]

One possible economic factor that may have played an important role in this respect was the fear among the relatively small-scale slaveholders of Fairfax County that their field hands would save their energy during the day in order to work in their garden plots at night, a situation they clearly could not afford considering their limited work forces, waning profit margins, and outstanding debts. Frederick Law Olmsted found that in those regions of the antebellum South where enslaved people were not universally permitted to cultivate garden plots, slaveholders were generally afraid that "the labor expended in this way ... tempts [slaves] to reserve for and to expend in the night-work the strength they want employed in their service during the day." One southern agricultural journal warned that slave families who were given their own garden plots "labor at night when they should be at rest."[14]

Indeed, when one considers the low productivity of slave labor in local wheat and corn cultivation, slaveholders' desire to intensify production at the expense of their slaves' garden privileges becomes more understandable. A rough calculation indicates that even on the relatively progressive plantation of Richard Marshall Scott in 1850, the estimated annual net return per able-bodied slave (between ten and fifty-five years old) employed in wheat and corn cultivation came to a meager $90.38, or about one-third the average return of enslaved people employed on rice plantations, and one-seventh the return of those working on sugar plantations (as shown later in the chapter).[15] Indeed, the loftier annual hiring rates for the county often ranged from $90 to $120 for able-bodied hands. In other words, slaveholders could sometimes earn more by hiring out their slaves by the year than by employing them in wheat and corn cultivation, which explains the unusual prevalence of long-term hiring in the region. The slaves owned by many of Scott's neighbors were even less productive, however. One local man was of the opinion that slaves in northern Virginia "have no value as field-hands" whatsoever, because "the labor performed by them is not sufficient to meet the current expenses of the plantations, at least the more ordinary ones." Slaveholders knew that their slaves' productivity was low, and they certainly would have had good reason to fear that if their field hands saved their energy during the day to cultivate garden plots at night, productivity would decline even more.[16]

This may explain why many of those who were permitted to cultivate garden plots were either elderly enslaved people, whose productivity on the farm had long passed its prime and whose provisions had been reduced accordingly, or especially industrious slaves, who set a good example for the rest. At Pohoke plantation the elderly bondsman Dick was given a small garden plot by his master after decades of faithful service. He gratefully told a visitor, "There is few masters like the Squire. He has allowed me to . . . take in a patch of land where I can raise corn and water Melions. I keep chickens and ducks, turkeys and geese, and his lady always gives me the price of the [Alexandria] market for my stock." This made economic sense. As Dick was no longer a productive member of the work force—he had been reduced to odd jobs such as milking cows and carting firewood—there was no real risk in a slowdown. Allowing Dick the privilege of partially feeding himself, moreover, may have eased his master's mind about issuing a faithful servant less than full rations. Those slaveholders who granted family garden privileges as rewards to especially industrious slaves probably calculated that such enticements would increase productivity among the rest of their able-bodied slaves. The above-mentioned Dick related that unlike the other slaves on his plantation, he had always done his work "without ever grumbling," which he no doubt felt made him more deserving of gardening privileges. Slaveholders thus maintained control over their slaves' internal production by either granting the right to garden plots or taking it away if productivity began to decline. Contrary to other slaveholding regions of the antebellum South, it seems that in Fairfax County family garden plots never took the form of a customary right of enslaved people.[17]

While slaveholders certainly acted in their own economic interest when it came to managing their human property, social factors may have further induced them to limit the development of slave family economies. Most southern planters—even those who allowed their slaves to cultivate family garden plots—felt uneasy about reward systems that increased their slaves' independence. They were especially reluctant to allow their slaves to act as proto-peasants, and many feared that enslaved people's possession of "too much money" would encourage bad habits (such as drinking and gambling) and disorder, not to mention add to their sense of power. In regions of the South where crops were more profitable and slavery as an institution was more tightly woven into the social fabric, slaveholders generally felt less threatened by their slaves' internal production. Slaveholders in Fairfax County, however, saw their social order crumbling, and to many the future of slavery seemed uncertain. In this context, slave behavior that smacked of independence may have caused nervous slaveholders to tighten their controls against internal production and other privileges. Elijah Fletcher, who resided for some time on the plantation of Thomson Mason, was told by the overseer that when it came to managing the slaves, "it would not do to indulge them."[18]

Ironically, tightened controls did not always work to the slaveholders' advantage. With few incentives or opportunities to engage in internal production, many enslaved people in Fairfax County resorted to theft, seizing their best opportunities to provide for themselves and their families. Elijah Fletcher wrote that local men and women in bondage "have very little to eat" and that "they will steal whatever they can get a hold of." Indeed, he opined, "who can blame the poor degraded objects?" One traveler in Fairfax County heard from a worried local that the enslaved population had very little access to additional food sources to supplement their weekly rations: "Speaking of the general condition of the slaves, he assured me that they were half-starved; to use his own words, they had hardly food enough to keep body and soul together. When they rob the hen-roost or the pasture field, it is to appease the cravings of hunger." One former slave from northern Virginia told interviewers that enslaved people on his farm "would visit the hog-pen, sheep-pen, and granaries—they were driven to it as a matter of necessity. . . . If colored men steal, it is because they are brought up to it." Charges of starvation were probably exaggerated, as local mortality rates were indeed lower than in other parts of the South (as demonstrated in chapter 3). For some, however, stealing to supplement standard rations and alleviate the pangs of hunger may have been legitimate. The above-mentioned slave further said that stolen goods could be sold to "poor white men, who live by plundering and stealing . . . paying in money, whiskey, or whatever the slaves want." In 1836, the citizens of Fairfax County even petitioned the General Assembly of Virginia for a law that would force local peddlers to obtain a special license from the county court, as such men were usually "persons not of good fame" who were eager to trade with the enslaved population for goods "plundered from our farms."[19]

Enslaved people in northern Virginia earned a stubborn reputation as petty thieves, which paradoxically served as yet another reason for slaveholders to deny them the privilege of cultivating family garden plots. Olmsted remarked that many planters suspected that the produce obtained from slaves' family garden plots could be used to "cover much plundering of their masters' crops, and of his live stock." This would have been relatively easy in Fairfax County, where agriculture was diversified and a number of foodstuffs were commercially cultivated on the farm that could also be produced in slaves' gardens. Incidences of "crop plundering" certainly occurred elsewhere in the region. One enslaved man living south of Fairfax County recalled that on his farm slave families often "cultivated a few square yards in corn . . . merely as a pretext for reaping a large crop." Sometimes, he added, "a daring theft would provoke a general search throughout the neighborhood." Though forbidden by law to sell any wares, Fairfax County slaves participated in a black market in stolen goods.[20]

Providing for Their Own Tables: Georgetown District, South Carolina

Very different was the situation on the plantations of the lowcountry. The task system, combined with crop-specific labor incentives offered by their masters, provided a powerful stimulus for enslaved families on the rice plantations to work for themselves, engage in internal production, and accumulate property. On the lowcountry rice plantations, enslaved people's family economies afforded them material conditions that were unparalleled in the antebellum South.

Whether or not enslaved men chose to voluntarily hire out their labor in exchange for cash, most of those who finished their tasks in the early afternoon certainly had the time to do so. Rice plantations also contained more skilled artisans—who found themselves in an advantageous position with respect to hiring out their labor—than the wheat and corn farms of northern Virginia. J. Motte Alston's plantation, for example, contained "carpenters, blacksmiths, coopers, tanners, shoemakers, tailors, bricklayers, etc." These men should have easily been able to convert their skills into financial gain. On Robert Allston's plantations, coopers received additional pay for extra work: twenty-five cents for every barrel over and above the daily task of three. Slave carpenters were also especially in demand, and some hired themselves out to perform work on other plantations or around the community as needed.[21]

Despite the amount of free time afforded them by task work, however, little evidence suggests that field hands in the region resorted to performing much outside day labor to earn extra cash for their families. Indeed, enslaved families on the rice plantations generally preferred to work together as collective economic units close to home. In this manner, they integrated valuable family time with cooperative efforts to improve their material conditions. They also maximized their internal production, procuring and cultivating a number of foodstuffs both for their own consumption and for sale.

Rather than hire out their labor individually in their spare time, enslaved men in Georgetown District often hunted and fished in groups that frequently included fathers and sons, brothers, or male members of extended families. J. Motte Alston of Woodbourne wrote that after completing their tasks on his plantation, enslaved men would "usually go to the sea shore and lay in a supply of fish and clams. Large numbers of mullet were caught at night in cast nets, and sacks full brought home." William Oliver, a former bondsman, recalled that during slavery men would "get pike out of the lakes" for their families to consume. They also hunted in the extensive forests that surrounded the slave village. "Game was all over the woods. Everybody could hunt everybody land those days. Hunting was free. . . . Wild turkey. Possum. Don't bother with no coon much." Evidence indicates that some rice planters actually permitted enslaved men to keep firearms with which to hunt game. Ben Horry told interviewers of the Federal Writers' Project that one particular creek in the cypress forest on his plantation was a "good place for go shoot squirrel. Give 'em name Squirrel Creek." He also recalled shooting birds: "Rice bird come jest as tick as dat. Sometimes a bushel one shot." Planters had good reason to give their slaves guns to shoot rice birds during the day because they threatened the crops. But slaves' possession of firearms was strictly illegal and it is unclear from Horry's testimony whether bondsmen on his plantation were allowed to bring guns home with them in the evenings. (They may have secretly hunted squirrels during the day.) In the lowcountry, however, permission for slaves to keep guns was certainly not unheard of. Olmsted was astonished to learn that the rice planter with whom he stayed on his visit to the Georgia lowcountry ignored the law and allowed his slaves "to possess, and keep in their cabins, guns and ammunition, for their own sport." Georgetown planter Plowden Weston, on the other hand, strictly forbade his slaves from keeping "any gun, powder, or shot," but the fact that he specified this rule so clearly suggests that guns may have been permitted by many of his peers.[22]

More important to slaves' internal economies were family garden plots. As historian Leslie Schwalm has argued, "both time and land were the crucial building blocks of internal production" for enslaved people. In Georgetown District, slaves had both. Unlike slaveholders in Fairfax County, rice planters allowed all of their slaves to cultivate family garden plots and raise poultry and even hogs, after their daily tasks had been completed.[23] They could afford to—the productivity of their slaves was independent of the number of hours they spent in the fields, so planters had little reason to fear a slowdown. Indeed, despite the fact that they spent altogether less time in the fields than slaves who worked under the gang labor system, the net returns of able-bodied field hands in the cultivation of rice were significantly higher than in most other regions. On Joshua John Ward's plantations in 1850, each able-bodied slave earned his master approximately $243.57 after expenses, as opposed to $90.38 per hand on R. M. Scott's relatively progressive farm in Fairfax County.[24]

Responding to this incentive to provide for their own tables and reduce their dependence on the master for sustenance, slave families in Georgetown District worked together to cultivate a wide variety of agricultural products in their spare time. Numerous sources confirm that slaves seized the opportunity to provide for themselves. Henry Brown, a former bondsman, remembered that slave families on his plantation "had all the vegetables they wanted; they grew them in the gardens." William Oliver likewise told interviewers that his family "had our garden. Different bean and collard. Turnip." Travelers to the district commented on the gardens of the slave population. One visitor noted that after their work for the master was finished, slave families "were at liberty to turn their attention to some little patch of ground, allowed them for their own particular use." Adam Hodgson remarked that besides their usual provisions of corn and sweet potato, slave families supplemented their diet "from the produce of their garden, and fish [which] they catch in the river." Captain Basil Hall observed that once a field hand had completed his daily task, he could "go home to work at his own piece of ground, or tend his pigs and poultry." And William Wyndham Malet, who visited Georgetown District in the summer of 1862, recorded that "the gardens here produce delicious figs, grapes, and melons, okra (what we call quash in India), egg plant, tomata—all in abundance. The negroes have all these in their gardens too."²⁵

The work and garden system afforded some slave families in Georgetown District the opportunity to achieve a degree of self-sufficiency almost unheard of in the slave South. Former bondswoman Hagar Brown recalled that her family grew so many vegetables in their garden plot that they were independent of their weekly corn rations, which they instead fed to the chickens: "Give you rations. Jess according [to how] many chillun you got. Ma say, 'Chillun, feed all the corn to the fowl.' Chillun sing, 'Papa love he fowl / Papa love he fowl / Three peck a day / Three peck a day.'" This was not uncommon. One visitor to antebellum Georgetown District reported that slave families were "often in a condition to sell the corn that is dealt out to them, having various means of providing for their own tables."²⁶

To maximize internal production, the cooperation of each family member was required in tending the family garden plot, hunting, fishing, and performing other domestic duties. As Hagar Brown's testimony makes clear, children were no exception. In her household, she and her brothers and sisters fed the chickens. Other enslaved children were expected to help out in the garden or otherwise contribute to the family economy—for example, by fishing. British reporter William Howard Russell observed one afternoon during his visit to Georgetown District: "the [negro] children of both sexes, scantily clad, were fishing in the canals and stagnant water, pulling out horrible little catfish," presumably to take home for consumption. Enslaved women, who, as elsewhere, were burdened with the responsibility of performing a number of domestic duties, did more than their share when it came to improving the material conditions of their families. Not only did they cultivate a variety of vegetables in their assigned garden plots and tend to their families' hogs and poultry, but they sometimes succeeded in converting their skills into money or bartered goods for the family. Margaret Bryant, a former bondswoman, recalled that her mother used to weave cloth to sell to local poor whites: "Po' buckra [poor whites] come there and buy cloth from my Ma. Buy three and four yard. Ma sell that, have to weave day and night to make up that cloth. . . . Don't have money fuh pay. Bring hog and such like as that to pay." In general, lowcountry slave families worked together as economic units and "pooled their efforts" to improve their standard of living. Larry Hudson has argued that "without assistance from family members, individual slaves would have struggled fully to exploit the land they tilled and other means of reducing their dependency [on their master]."²⁷

The drivers, who often found themselves at a disadvantage with respect to internal production due to their longer workdays, were in some measure compensated. Many drivers were permitted to demand assistance from other slaves in planting and cultivating vegetables in their garden plots, as research by Philip Morgan has shown. One former bondsman from the lowcountry recalled that the "drivers had the privilege of planting two or three acres of rice and some corn and having it worked by the slaves." According to Leslie Schwalm, drivers on the rice plantations were rewarded with "better housing, better housewares, improved rations, greater gardening privileges, and exclusive access to milk cows." Rice planter William Lowndes allowed his drivers the exclusive privilege of planting rice for sale (to him). He noted in his account book that "each Driver saved his own seed. Ben's stacks turned out to be 43 bushels rough rice. Hickory's and Philip's stacks each about 35 bushels." Nevertheless, some drivers felt compelled to steal from their owners to acquire small luxuries. Ben Horry, the son of a driver on Joshua John Ward's Brookgreen plantation, related the following story to interviewers: "My father the head man, he tote the barn key. . . . He ain't have money but he have the rice barn key and rice been money! . . . I been old enough to go in the woods with my father and hold a lightard (lightwood) torch for him to see to pestle off that golden rice he been tote out the barn and hide. That rice been take to town Saturday [to sell]." In this particular case, Horry's father did not sell the stolen rice to buy any gifts for his family—"With the money he get when he sell that rice, he buy liquor," he added—but the testimony suggests that some drivers were not sufficiently compensated for their long hours working for the master.[28]

The development of slave family economies as a system of both monetary exchange and bartering was greatly stimulated by the role that rice planters and their families played in providing their bondspeople with a market for their surplus wares and produce, which they usually bought themselves. Historian Charles Joyner argued in his study of the antebellum slave community on Waccamaw Neck in Georgetown District that it was common for rice planters to "purchase hogs, poultry, produce, or cord wood from [their slaves]." The account books of Robert F. W. Allston reveal that he regularly gave slave families breeding hogs, the offspring of which he purchased each year at the price of $5 per hundred pounds. Travelers also commented on such practices. William Howard Russell noted that throughout Georgetown District, "the negroes rear domestic birds of all kinds, and sell eggs and poultry to their masters." Swedish traveler Fredrika

Bremer observed that slave families living on the Pee Dee "sell their eggs and chickens, and every Christmas their pig also, and thus obtain a little money.... They often lay up money; and I have heard speak of slaves who possess several hundred dollars." Georgetown planter James Sparkman also spent a significant amount of money on purchases of meat and produce from his slaves. He boasted to a fellow planter: "As an illustration of the indulgences which my own people enjoy, I have during the past year kept an item of their perquisites from the sale (to me) of Eggs, poultry, Provisions saved from their allowance, and the raising of hogs, and it amounts to upwards of $130." Sparkman was not only his slaves' primary customer; he was also their banker: "They frequently ask me to become Treasurer of their little funds." Olmsted reported similar practices in the Georgia lowcountry.[29]

Slaveholders bought their slaves' goods and became "treasurers" of their funds primarily as a way of controlling their access to money and preventing them from peddling their wares to outside parties beyond the slaveholders' control for goods they did not want their slaves to have (such as alcohol). However, although the practice was forbidden, enslaved people sometimes bypassed their masters and flouted the law by trading their surplus goods to customers off the plantation when it was profitable to do so. In so doing, they created opportunities to secure the highest possible returns for their products. The mother of Margaret Bryant wove cloth to sell to "po' buckra," as stated above. J. Motte Alston learned one day that the wife of a poor white man "had traded with my negroes, which offense, by the laws of the State, was punishable by whipping at the 'Carts-tail' on the public square of the district Court-house." Because it was a woman, however, Alston chose not to alert the authorities. William Howard Russell learned that local slaves found a lucrative market in selling their wares to the crews of the riverboats that frequently criss-crossed the Winyah Bay area. Discussing the matter with one such ship's captain, he learned that the captain "can buy enough of pork from the slaves on one plantation to last his ship's crew for the whole winter. The money goes to them, as the hogs are their own." Slaves' determination to seek maximum profit for their goods placed them in an advantageous bargaining position vis-à-vis their masters, who were forced to offer the most attractive prices to their slaves in order to combat illicit trade with poor whites. James Sparkman claimed that his slaves "get better bargains [by selling their goods to me] than can be made by themselves with the shop keepers." Robert F. W. Allston likewise offered good money for his slaves' hogs, warning that if any hogs were sold off the plantation, the guilty slave would forfeit his or her privilege of keeping animals.[30]

Rice planters acted in their own interest when they offered to buy their slaves' produce, but they stood to benefit from the work and garden system in other ways as well. First, masters who lured their slaves with the prospect of engaging in internal production during their free time could rest assured that every aspect of plantation work would be accomplished in a timely manner. Enslaved field hands worked quickly and efficiently on the rice plantations in Georgetown District because they were highly motivated to take advantage of the opportunity to work for themselves and their families, thereby strengthening the work ethic in the eyes of planters. Larry Hudson has argued, moreover, that lowcountry planters who employed the work and garden system hoped to "instill in the slave a sense of industry."[31]

Second, planters who provided their slaves with the opportunity to cultivate family garden plots aimed to encourage, according to Hudson, "the slave literally to 'grab a stake in slavery'; the master was likely to get the best out of workers who felt some attachment to the home place, and, by extension, to the master." In other words, by offering slaves patches of land for family garden plots, as well as the means to convert surplus goods into hard cash, planters sought to further bind their workers to the plantation and discourage flight. As one Carolina overseer aptly put it, "No Negro with a well stocked poultry house, a small crop advancing, a canoe partly finished, or a few cubs unsold, all of which he calculates soon to enjoy, will ever run away." James Sparkman concurred. Claiming that "comforts have multiplied" among the slaves on local rice plantations, he found that "runaways are fewer and less lawless."[32]

The amount of free time afforded enslaved field hands under the task system is a third reason that nervous rice planters encouraged their slaves to engage in internal production. Indeed, planters were convinced that slaves should be kept productively occupied during their relatively ample leisure time to keep them out of mischief. One planter felt that slaves should always be given garden plots to cultivate during their "idle hours," which might "otherwise be spent in the perpetration of some act that would subject them to severe punishment." And finally, as Leslie Schwalm has argued, planters aimed to transfer at least some of the burdens of provisioning by encouraging a degree of self-sufficiency among their slaves. The knowledge that bondspeople would supplement whatever rations they received with a variety of meats and vegetables allowed at least some rice planters to distribute the bare minimum to their workforce without damaging their self-image as benevolent paternalist masters. Robert F. W. Allston, for one, supplied his slaves with only what he considered "absolutely necessary for [their] health and endurance." (Not all planters were so inclined, however. Plowden Weston was adamant that his slaves receive plentiful rations: "in all cases of doubt, [rations] should be given in favour of the largest quantity.")[33]

Despite the underlying reasons for such incentives, lowcountry slaves seized every opportunity to make life for themselves and their family members more comfortable. Enslaved people on Birdfield plantation, for example, spent over $110 in 1858 on "comforts and presents to their families," including "Sugar, Molasses, Flour, Coffee, Handkerchiefs, Aprons, Homespun and Calico, Pavilion Gause (Mosquito Nets), Tin Buckets, hats, pocket knives, sieves, etc." The material conditions enjoyed by slaves in the lowcountry never ceased to amaze visitors. Discussing the matter with the captain of a Winyah Bay ferry, William Howard Russell learned that "the Negroes on the river plantations are very well off. . . . One of the stewards on board [the riverboat] had bought himself and his family out of bondage with his earnings." Olmsted reported that slaves' cabins in the lowcountry all had locks to protect their valuable belongings from theft, and even observed a slave family driving their own horse and carriage to church. Clearly, task work and labor incentives afforded enslaved people in the lowcountry numerous attractive opportunities to improve their material conditions.[34]

Against the Odds: St. James Parish, Louisiana

The specific demands of sugar cultivation in St. James Parish had far-reaching consequences for enslaved families' opportunities to work for themselves and develop family-based independent economies. It goes without saying that slaves living on the sugar plantations of the Louisiana cane world had the least amount of free time of slaves living in any other agricultural region of the nineteenth-century South. Yet, remarkably, although slave families lacked much free time, evidence indicates that they did in fact manage to develop moderate family-based economies—more developed than the thwarted family economies of Fairfax County, but less developed than the extensive internal economies of slave families living in Georgetown District. The most significant factor in this respect was the prevalence of various crop-specific labor incentives offered by their masters. Whereas in Fairfax County many slaves lacked both the time and the means to develop extensive family-based economies, and slaves in Georgetown District had both time and the means, in St. James Parish slave families lacked much time but were overwhelmed by the means to improve their material conditions.

Paid overwork in the sugar country provided many enslaved people, especially men, with opportunities to earn extra cash for themselves and their families. Unlike in Fairfax County, such opportunities in St. James Parish were plentiful. The industrial nature of sugar production, for instance, required a wide variety of skilled slaves—including bricklayers, carpenters, coopers, engineers, mechanics, sugar makers, and blacksmiths—many of whom were in a position to occasionally profit from their valuable skills and earn money to buy luxury goods for their households. A slave carpenter on Houmas plantation named Peter, for example, earned $5.00 for reframing the windows and doors of a neighboring overseer's house. Although many slaveholders instructed their overseers to "not allow the Mechanics to make or sell any of their work without special permission," most paid their own skilled slaves for extra work, especially during the grinding season. On the Bruce, Seddon, and Wilkins plantation in St. James Parish, a slave carpenter named Anderson was paid regularly for extra work. On the same estate during the grinding season of 1849 the enslaved sugar makers were each paid a bonus of $7.50 for their shifts in the sugar house. Planter Benjamin Tureaud paid his slaves for making bricks and hogsheads, while others paid their slaves for making rails and handbarrows during their time off.[35]

Not only skilled slaves were afforded the opportunity to engage in paid overwork, however; a vast majority of the paid overwork performed on the sugar plantations of St. James Parish was done on Sundays by ordinary field hands. Usually there was so much work to do on a sugar plantation that sugar planters welcomed extra work in return for small payments in cash. Chapter 2 noted that many slaves in St. James Parish worked on Sundays; evidence suggests, however, that most were financially compensated for it. Victor Tixier, during his travels through St. James Parish, found that "Sunday belongs to [the slave]; then he works only for money." Eliza Ripley, the daughter of a local sugar planter, recalled after the Civil War that "from Saturday noon till Monday was holiday, when the enterprising men chopped wood, for which they were paid." As Ripley's testimony makes clear, paid overwork on the sugar plantations of St. James Parish revolved mainly around chopping wood—procuring extra fuel for the boilers, kettles, and steam engines of the sugar house. Some three to four cords of wood were needed to produce every one hogshead of sugar in nineteenth-century Louisiana. For large planters like Valcour Aime, who produced over a thousand hogsheads of granulated sugar in 1853, the amount of fuel necessary to keep the sugar-making machinery operating during the entire grinding season was substantial to say the least. There was almost never enough time during the normal workweek for slaves to procure enough fuel for the grinding season, and so sugar planters regularly enticed their slaves with monetary rewards to perform such work on Sundays and other holidays.[36]

Nineteenth-century plantation journals for St. James Parish abound with accounts of slaves chopping or collecting wood on Sundays, suggesting that most enslaved men in the region regularly availed themselves of the opportunity to earn extra cash, even if it meant sacrificing their one free day in the week. In some cases it appears that fathers and sons or other family members worked together, as they were paid together in lump sums. On Benjamin Tureaud's plantation, for example, pairs of slave men occasionally received payment for wood collected in 1854. Michel and Isaac, Daniel and Thomas, William and Philip, and the "Butcher Family" are just a few examples. Tureaud paid dozens of others for collecting wood as well. One man named Frank earned $8.80 for the driftwood that he collected at the river. Ben Runt earned $5.00. Smaller payments were more common, however. Isaac Faulk's case was fairly typical. On one Sunday he earned $0.40, and on another occasion $0.60 for catching driftwood. On occasion wood was even collected by elderly women or small children on Tureaud's estate. Little Joseph, a child, was able to earn $1.60 in 1858 for helping to collect wood. Old Catherine earned $0.70. Sugar planter Octave Colomb likewise paid his slaves substantial sums for cordwood. Again, family members working together were not uncommon. John and Jim Barney earned $6.87½ for 13¾ cords of wood in 1850. Clark, Tom, and C. Cooper collectively earned $10.00 for procuring twenty cords of wood. In total on Colomb's plantation, thirty-four men collected wood on Sundays in 1850, collectively earning some $117.25. So it was throughout the parish. Some planters paid enormous sums for cordwood. Slaves on the Wilton plantation collectively earned $53.65 on a single day in 1849. On Uncle Sam plantation, fifty-three slaves received a total of $600 for cordwood in 1860.[37]

Procuring fuel for the grinding season was not the only way to earn extra cash through paid overwork. Seasonal demands encouraged slaveholders to pay their slaves to perform a variety of other jobs as well. T. B. Thorpe, a visitor to one of the largest sugar plantations in the parish, noted that during the Christmas holidays the master "pays for innumerable things, which have been provided by the slave, without interfering with his accustomed labors." Octave Colomb paid his slaves to work "on the pond" on Sunday, 5 September 1852. The Wilton plantation slaves were occasionally paid for ditching and levying, including on Christmas Day, 1849. Three enterprising women on Tureaud's estate earned substantial sums for unspecified overwork in 1858. A certain Anna earned $28.00 for fifty-six days of paid overwork—practically every Sunday and holiday she had. Sarah Dom earned $22.00 for forty-four days' work, and Big Winny earned $24.00 for forty-eight days' work. Such extreme cases were rare, however, as few were willing to give up virtually all of their free days to perform paid overwork.[38]

Collecting and selling the Spanish moss that grew in abundance on the plantations of southern Louisiana provided slaves with yet another lucrative and simple way of earning extra money. One visitor remarked during a journey through the sugar parishes that the moss that gracefully draped the live oaks in the region had "become a considerable article of commerce . . . used to stuff sofas and mattresses." Slaves were keenly aware of its commercial potential, and most sugar planters permitted their bondspeople to collect and sell the moss that grew on their estates. One former bondsman remembered in an interview with workers of the Federal Writers' Project, "Once I heard some men talkin'. One said, 'You think money grows on trees?' The other one say, 'It do. Get down that moss and convert it into money.' I got to thinkin', and sure 'nuff, it do grow on trees!" If not sold, moss could be used in the slaves' own houses. William Howard Russell noticed that the slaves on Governor Roman's plantation in St. James Parish slept on mattresses stuffed with Spanish moss they probably collected themselves.[39]

Not only opportunities for paid overwork but also patches of land were made available to enslaved families in St. James Parish. As in Georgetown District, slave families on most of the sugar plantations of St. James Parish were permitted to cultivate small garden plots and to raise poultry and hogs, both for their own consumption and for sale. Again, most sugar planters could afford to grant their slaves such privileges, as productivity on the sugar plantations was exceptionally high.

T. B. Thorpe's statistical account of the plantation he visited in 1852 (he kept the owner's name anonymous) provides an interesting example of slave productivity at the high end of the spectrum. In 1852, his host owned some 215 slaves, 107 of whom were able-bodied field hands. During that exceptionally successful year, these slaves earned their master approximately $900.84 each, after subtracting average expenses for clothing and food.

Even if other expenses are considered, such as repairs of machinery and buildings, as well as the salaries of overseers and engineers (which according to Thorpe came to around $20,000 per year), this still would have left the master with $743.93 profit per hand. Thorpe's host may have been one of the most progressive and successful planters of the parish, and 1852 may have been an exceptional year, but despite wide fluctuations in harvests and profits, slave productivity generally remained very high throughout the antebellum period. As a result, slaveholders had little reason to fear substantial losses by allowing their slaves to cultivate family garden plots. Moreover, as in the lowcountry, sugar planters themselves benefited from garden privileges (as discussed below).[40]

Visitors to the Louisiana sugar country frequently commented on the presence of vegetable gardens and animals around the slave quarters. Joseph Holt Ingraham remarked during a journey to the region in 1835 that "to every [slave cabin] was attached an enclosed piece of ground, apparently for a vegetable garden." Francis and Theresa Pulzsky likewise observed that "the negroes have their own little gardens, they keep their poultry and sell it to their master." They further noted that some slaves were occasionally even permitted to sell their goods "to the steamboats" that regularly passed by, "but never without a written '*permit*.'" T. B. Thorpe wrote during his visit to a local plantation that "in the rear of each cottage, surrounded by a rude fence, you find a garden in more or less order, according to the industrious habits of the proprietor. In all you notice that the chicken-house seems to be in excellent condition." Olmsted found that slave families living on the sugar plantations he visited raised "corn, potatoes, and pumpkins" in their garden plots, which they could either sell or consume themselves, although "generally they sold it." He observed furthermore that the slaves "worked at night, and on Sundays on their patches, and after the sugar and corn crops of the plantation were 'laid by,' [their] master allowed them to have Saturday afternoons to work their own crops."[41]

Enslaved families living on the sugar plantations worked hard to supplement their weekly rations. Their limited amount of free time, however, reduced them to raising vegetables and keeping animals that required a minimum amount of time and attention. Pigs and poultry, when permitted, were easily raised. On the plantation where bondsman Alexander Kenner lived in neighboring Ascension Parish, slaves were known to "raise a thousand dozen chickens in a year," and the master obliged them to "sell the chickens to him, instead of selling them to hucksters." William Howard Russell, visiting a St. James Parish sugar plantation, noticed that in the yards surrounding the slave cabins a great number of "pigs and poultry were recreating." Likewise, slave families cultivated vegetables that could be grown without too much effort. Especially popular, as Olmsted observed, were vegetables such as corn, potatoes, and pumpkins because their cultivation in gardens required a minimum amount of labor and attention. Former slave Francis Doby recalled that her family also grew "turnips, carrots, [and] cabbage—all a-comin' from de patch."[42]

Planters especially encouraged their slaves to grow corn since it formed the basis of the slaves' diet. Indeed, sugar planters in St. James Parish even allotted slave families extensive patches of land known as "negro grounds" upon which primarily corn, and to a lesser extent pumpkins, could be cultivated. These provision grounds were given to slave families *in addition* to their garden plots. They were usually located along the peripheries of the plantation, often at a great distance from the slave village. One visitor to John Burnside's plantation observed that "on the borders of the forest the negroes were allowed to plant corn for their own use." These plots could only be cultivated on Sundays, but sugar planters generally afforded their slaves a few days off in the spring and again in the fall, before the grinding season commenced, to plant and harvest their corn crops. Valcour Aime recorded in his plantation diary in October of 1842: "Gathering the corn crop of the plantation hands on 12th, 13th, 14th and 15th. Some hands raised thirty-six barrels of corn to the arpent." While slave families consumed what they produced in their vegetable gardens, they generally sold to their masters the corn they grew on their additional provision grounds (and the masters in turn dealt it back out to the slaves as rations). Indeed, corn sales were—next to wood chopping—one of the major sources of additional income for enslaved families. On Benjamin Tureaud's plantation, David Rock's family earned $67.25 for corn in 1858. George Cane's family earned $27.70. Big Mathilda was paid $10.00 for the 700 pumpkins that she and her family grew on their provision grounds. Ex-slave interviews also testify to the importance of corn as a source of income. Former bondswoman Elizabeth Ross Hite remembered that her "mother planted corn, but de master bought it from her. He paid fifty cents per barrel for corn."[43]

The development of family economies among the enslaved was stimulated by high demand for their products, and in St. James Parish families were afforded the opportunity to market their goods right on the plantation. Plantations in southern Louisiana frequently contained commissaries where money and goods obtained from slave families' internal production could be exchanged for small luxuries. The records left by nineteenth-century sugar planters in St. James Parish abound with slave family accounts at the plantation stores, offering an interesting glimpse into the consumer behavior of bondspeople. Purchases made by slaves were entered into the account books as debts, while products sold to the store—such as corn or cordwood—were entered as credits. These accounts were listed under the names of the heads of families, but women and children could and did make purchases under their husbands' and fathers' accounts. Most enslaved people spent their hard-earned money on small luxuries for both themselves and their family members.[44]

On Benjamin Tureaud's estate, for example, a slave named George Cane, whose account was credited from wood and corn, bought his wife a dress for $2.00 on 8 August 1858. Another man named Sampson likewise bought his wife a dress for $2.50. Willis's account was credited that same year with the exorbitant sum of $67.55 from wood and corn. His purchases included a pair of children's shoes for $0.55, adult shoes, a shirt, twenty-five pounds of meat, flour, and several pounds of tobacco. One slave's account was credited with $3.00 for the sale of twelve baskets, which were subsequently sold in the commissary to fellow bondsmen on the plantation. Other articles sold to the slaves included the following: locks, tin, hose, check fabric, handkerchiefs, pants, calico, thread, buckets, rice, linen, and shirts. On the Bruce, Seddon, and Wilkins plantation a similar system was in use. Several families' accounts were credited for the sale of eggs and chickens, cordwood, corn, and pumpkins. With their earnings, John Lewis and his family purchased a basket, a pair of suspenders, several bars of soap, a knife, a hat, a pipe, and several plugs of tobacco at the plantation store in 1848. One man named Luke bought a pair of shoes for $1.50 for his wife, Priscilla. Most purchases made by slave families at plantation stores throughout the parish were modest, however. As a result of the limited amount of free time they had to engage in internal production, slave families were generally unable to accumulate a great deal of property in St. James Parish.[45]

Less frequently, slave families marketed their goods to third parties, even though this practice was strictly forbidden. When Octave Johnson ran away to a forest located four miles to the rear of his sugar plantation, where he remained truant for a year and a half, he survived partly by hunting and secretly trading meat for cornmeal from the field hands. This was an exceptional case, but there was enough evidence of slaves trading with river traders to alarm local slaveholders, who built commissaries on their estates partly as a means of thwarting any traffic that could occur beyond their control. William Kingsford, who visited the sugar country in 1858, claimed that river peddlers "pass from plantation to plantation, trading with the negroes principally, taking in exchange the articles which they raise, or, when the latter are sold to the boats, offering to their owners the only temptations on which their money can be spent." Victor Tixier reported that in St. James Parish, the slave villages are "built at a great distance from the river, in order that the rivermen of the Mississippi, this plague of Louisiana, cannot have traffic with the Negroes at night." The opportunity to market their products to third parties offered daring slaves the chance to obtain higher prices for their goods or to purchase articles, such as whiskey, that were not available from plantation stores. More important, they encouraged slaveholders to offer their slaves attractive prices for their products to discourage illegal trafficking, as in Georgetown District.[46]

While slave families in St. James Parish eagerly took advantage of the incentives offered to them by their masters to engage in internal production, often sacrificing their only free day in the week to do so, sugar planters profited as well from their slaves' willingness to work for themselves. As in Georgetown District, the economic benefits of such incentives outweighed slaveholders' social distaste for the "independent behavior" of their slaves. Sugar planters very well realized that the particularly taxing nature of sugar cultivation, and especially the lack of free time it permitted slaves, could have negative consequences for slaves' productivity and morale. As Ira Berlin has noted, "in trespassing on their slaves' free Sunday or half-Saturday, [sugar] planters found themselves in dangerous territory as they risked further alienating and demoralizing their workers." Local sugar planters preferred to play it safe and "entice [their slaves] to the field" with a number of attractive labor incentives, which they modeled after colonial precedents. The slave conspiracy at Pointe Coupée in the 1790s had convinced Spanish colonial authorities to legally require slaveholders to offer their slaves family garden plots, compensation for Sunday work, and adequate food and clothing. Such measures were designed to keep slaves content so that they should "banish from their minds the notion of acquiring liberty," a cause with which American sugar planters in the nineteenth century could readily identify.[47]

Such incentives not only appeased their laborers but benefited planters in other ways as well. By offering to pay their slaves for overwork, for example, planters secured even more of their slaves' labor at a minimal cost and encouraged them to take an interest in the success of the plantation. Olmsted even learned of cases of limited profit-sharing among sugar planters and slaves during the exhausting grinding season: "At Christmas, a sum of money, equal to one dollar for each hogshead of sugar made one the plantation, was divided among the negroes.... It was usually given to the heads of families." He further noted that sugar planters believed in paying their slaves for extra work because it gave "the laborers a direct interest in the economical direction of their labor: the advantage of it was said to be evident." Moreover, planters in St. James Parish saved money by buying produce and cordwood from their slaves at below market prices. The latter was crucial for the success of the plantation, but the former also provided planters with substantial savings. By paying their slaves to produce the corn that was eventually ground up and rationed back out to slave families, for example, slaveholders were able to circumvent the hassles of contracting with merchants to secure their slaves' provisions and avoided having to divert too much of their slaves' efforts to cultivating provision crops during working hours, which they used to extend work in the cane fields instead. Some planters even resold their slaves' produce at a profit. Duncan Kenner, a sugar planter from Ascension Parish, bought chickens from his slaves at a rate of $0.20 each, only to resell them for $0.30. And finally, slaveholders discouraged illegal trading and controlled their slaves' access to money by encouraging—theoretically compelling—them to make their purchases at the plantation commissary. Paradoxically, as Richard Follett has argued, the labor incentives offered by planters "proved sadly profitable for the sugar masters, who rewrote the mutual obligations of paternalism to exploit their slaves still further."[48]

Conclusion

In different cash-crop regions of the nineteenth-century South, the specific nature of commercial agriculture resulted in distinct types of internal economies among slave families. In all three of the regions discussed here, the development of slave families' internal economies was steered primarily by the pressing demands of their formal work and both the financial and social interests of their masters. External factors denied slave families living in Fairfax County of both the time and the means to develop extensive internal economies. The boundaries encountered by slave families there were often impenetrable, and opportunities to work for themselves or negotiate for better material conditions were rare. In Georgetown District, bondspeople were afforded both the time and the means to work for themselves, offering slave families attractive opportunities to partially provide for themselves. In St. James Parish, slave families lacked much free time, but remarkably they took advantage of various labor incentives to work for their own gain on Sundays and other holidays.

The development of family-based internal economies in different cash-crop regions depended upon the specific external factors of local slave-based agriculture, suggesting that slaves' economic behavior was not of a "proto-peasant" nature. Indeed, nowhere did slave families even truly own the means of their internal production. Permission to cultivate garden plots, for example, was granted to slaves only when it served the interests of the slaveholders. Where it did not—or *appeared* not to, such as in Fairfax County—this privilege was denied slave families. The opportunity to engage in paid overwork was likewise determined by slaveholders and the demands of regional agriculture. Slave families were also seldom permitted to freely market the goods they produced from their informal labor. In Fairfax County the prohibition of such trafficking was strictly enforced, while in South Carolina and Louisiana the internal economy was designed by slaveholders to develop within a closed system in which slaves' goods were sold directly to the masters and not to third parties. To be sure, slave families' remarkable determination to work to improve their own material conditions is a testament to human resilience in the face of oppression. To describe slave families' internal economies as "autonomous," however, is in fact a misnomer. They were not autonomous but rather were circumscribed by, and dependent upon, the nature of their formal labor for the master. As one former slave from Louisiana tellingly put it, slaves could work for their own gain, but "we could not allow our work . . . to interfere with master's work."[49]

III

Social Landscapes

Family Structure and Stability

5

Slaveholding across Time and Space

The land and its products certainly played a guiding role in the daily experiences of slave families, during both their time for the master and their time for themselves, but at a more fundamental, demographic level the nature of regional agriculture also determined the very basis for family life. Local economic trends directly and indirectly influenced slaves' marriage strategies and the stability of their family relationships across time and space. To analyze slaves' family formation and experiences with forced separation in the coming chapters, particular attention must first be paid to the very nature of their containment in different parts of the South.

Factors such as the spatial distribution and sexual composition of enslaved populations, both of which defined the social landscapes in which they lived, were crucial in this regard. First, the spatial distribution of enslaved populations among slaveholdings of various size played an important role in determining the nature of social contact and marriage strategies among slaves. Largely restricted in space and mobility to the agricultural units upon which they lived and worked, bondspeople throughout the South turned first to the slave communities of their home residence for social contact, although slaveholders' frequent permission for weekend visiting generally expanded social networking in all but the most isolated regions. Second, the sexual composition of enslaved populations determined the physical possibility and extent of slave family formation in different localities. The boundaries and opportunities with which slaves were confronted when seeking a mate were thus greatly influenced by slaveholding size and sex ratios. These factors in turn varied across time and space, as dictated by the nature of slave-based agriculture in various communities of the nineteenth-century South.

This chapter broadly examines the nineteenth-century evolution of slaveholdings on the grain farms of Fairfax County, the rice plantations of

Georgetown District, and the sugar plantations of St. James Parish, respectively. What was the spatial distribution and sexual composition of enslaved populations in different regions of the non-cotton South, and how did they change over time? The aim of this chapter is to provide a basis from which to further explore enslaved people's experiences with family formation and stability in chapters 6 and 7 by first establishing the social landscapes of slave populations in each region.

The Downward Spiral: Fairfax County, Virginia

Just before the turn of the nineteenth century, Fairfax County ironically boasted one of the largest slaveholdings of the newly formed United States. Only a handful of eighteenth-century plantations were of a truly grand scale to begin with there, but George Washington's famous Mount Vernon, situated on the banks of the Potomac River, contained close to three hundred slaves upon his death in 1799. By the outbreak of the Civil War in 1861, however, the county no longer had even a single plantation of any grandeur. In Fairfax, the transformation of local agriculture and the economic hardships that plagued the county throughout most of the first half of the nineteenth century were paralleled by a spectacular decline in slaveholding size.[1]

Table 5.1 illustrates the general trend in Fairfax County. As early as 1810 some 58 percent of the slave population lived on farms (units with twenty slaves or fewer) rather than plantations (units with over twenty slaves), and only two slaveholdings—accounting for just 6 percent of the total slave population—contained more than one hundred slaves. By the eve of the Civil War, however, a staggering 84 percent of the enslaved population lived

Table 5.1. Distribution of the Fairfax County Slave Population by Slaveholding Size, 1810–1860

Year	Total Slaves	1–10 Slaves		11–20 Slaves		21–40 Slaves	
		#	%	#	%	#	%
1810	5,927	1,843	31	1,606	27	1,214	21
1820	4,770	1,959	41	1,248	26	819	17
1830	3,992	1,620	41	1,148	29	756	19
1840	3,440	1,721	50	949	28	554	16
1850	3,178	1,913	60	754	24	411	13
1860	3,116	1,796	58	808	26	469	15

Source: U.S. Census, 1810–1860, National Archives and Records Administration.

on small farms—58 percent lived on farms containing ten slaves or fewer—and only 16 percent were held in units of more than twenty. Virginia Scott's Bush Hill, by then the largest slaveholding in the county, contained only forty-three slaves, most of whom she annually hired out. The next largest slaveholder was Alexander Grigsby, who was listed in the census as the owner of forty slaves. Grigsby was a known slave trader, however; presumably he intended to sell most of his slaves to the Deep South. Not only had the size of Fairfax County slaveholdings reached an all-time low by 1860, but so had the size of the slave population itself. Indeed, between 1810 and 1860 the slave population was slashed by 47 percent, from 5,927 to just 3,116.[2]

At least three interrelated factors contributed to the general decline in slaveholding size in Fairfax County during the first half of the nineteenth century. First, excessive estate divisions obviously took their toll on the distribution of the slave population. Primogeniture law, by which a slaveholder's inheritance was passed on in bulk to the eldest son, was no longer practiced in Virginia by the turn of the nineteenth century and indeed was even abolished. (It was viewed by many as an outdated and unrepublican practice.) This meant that by the beginning of the nineteenth century, slaveholdings were being divided more or less equally among several heirs, effectively breaking up the colonial pattern of moderate and large slaveholdings.[3]

The breakup of these slaveholdings was exacerbated by the economic difficulties facing planters and farmers in the nineteenth century, which not only prevented expansion in either land or slaves but also triggered a decrease in slaveholding size. In other words, small slaveholdings tended to remain small—virtually all actually shrank over time. As a "distressed

41–60 Slaves		61–80 Slaves		81–100 Slaves		101–200 Slaves		+200 Slaves	
#	%	#	%	#	%	#	%	#	%
428	7	361	6	100	2	140	2	235	4
344	7	66	1	176	4	158	3	0	0
300	7	0	0	168	4	0	0	0	0
140	4	76	2	0	0	0	0	0	0
100	3	0	0	0	0	0	0	0	0
43	1	0	0	0	0	0	0	0	0

community oppressed by debt," as county residents collectively described themselves, the Fairfax County slaveholding class could not afford to expand its slaveholdings. Indeed, one visitor observed in 1835 that "the land-holders in [these] parts of Virginia are becoming poorer nearly in direct proportion to the number of their slaves." Quick to realize this, local slaveholders not only culled their naturally growing slave population but also reduced it in absolute numbers by almost half in the antebellum period, occasionally by manumission but mostly through sale to the Deep South via the domestic slave trade, where the demand for slave labor was practically insatiable. Farmers and planters tried to relieve some of their financial problems by making do with a minimum amount of slave labor; in other words, fewer slaves became responsible for more tasks, as discussed in chapter 2.[4]

A third important factor that contributed to the decline in local slaveholding size was the decreasing fertility of the soil, which triggered a shift from extensive to intensive cultivation in local agriculture. In other words, slaveholders expressly turned to cultivating smaller plots more intensively. Plantations devolved into farms, and farms devolved into smaller farms. One historian noted that in the nineteenth century, Fairfax County farmers were keen to "intensively cultivate smaller plots, without the hindrance of a 'plantation mentality.'" James Redpath, a northern reporter who traveled through Fairfax County in the 1850s, observed that by that time the land was "chiefly held in small sections." When Lawrence Lewis's unusually large Woodlawn plantation was offered for sale in 1846, the advertisement stated that "it can be divided into small farms." Farmers found that their scanty supply of manure, crucial in any attempt to revitalize the soil, could be spread over only a limited space. The least productive fields of many planters' inherited estates were therefore abandoned and left for the forests to reclaim. As the amount of actual acreage that was brought under cultivation declined, slaveholders could not only not *afford* more slaves, but they did not *need* more slaves. Redpath heard from a local farmer that it took only "two men and a boy to cultivate . . . twenty-five acres and attend to the cows." Most agreed that it was better to reduce their number of slaves to the bare minimum—extra hands (whether free or slave) could always be temporarily hired on during especially labor-intensive seasons such as the harvest—than to keep and maintain superfluous slaves. John Taylor, a northern Virginia planter and agriculturalist, warned his neighbors against keeping more slaves than could be personally supervised or put to efficient use. Other progressive farmers, such as Edmund Ruffin, also encouraged a reduction in the number of slaves and the procurement of better results from the remainder.[5]

This trend was given added impulse throughout the nineteenth century by the widespread adoption of more labor-saving devices, which further reduced the number of hands necessary to cultivate grains. Implements such as improved plows (which diminished part of the need for excessive hoeing), scythes, cradles, and harrows were popular among local farmers. Richard Marshall Scott, who in 1846 complained that it "will require much constant labor" to turn a profit from his inherited plantation, took a step in the right direction when he "went to town and bought a thrashing machine" in November 1848. The introduction of reapers in the wheat harvest resulted, according to one local farmer, in "the saving of much expense and great risk; by their aid with seven hands and two boys, fifteen acres can with ease be cut and shocked per day—the Self Rakers requiring one less hand." Dennis Johnston's estate, which was sold upon his death in 1853, contained "ploughs, harrows, wheat fans, [a] threshing machine, scythes, cradles, mowing scythes, forks, rakes, &c."[6]

Enslaved people in Fairfax County thus increasingly found themselves living on small holdings between 1800 and 1860, as local farmers and planters inherited small fractions of their forefathers' once moderate-sized plantations, and moreover they found that they could neither afford, nor did they need, many slaves. By the eve of the Civil War, a majority of local slaves lived on units with ten slaves or fewer. The precise distribution of the enslaved population in nineteenth-century Fairfax County is more difficult to ascertain, however, as locating the farms and plantations listed in the census schedules is not always possible. The county was divided into two parishes: Truro, which was lower lying and bordered the Potomac, and Fairfax, which consisted of the upper western and interior sections. In his demographic study of the Fairfax County slave population, historian Donald Sweig found that throughout the antebellum period a majority of local slaves were "clustered" in the lower parish of Truro, probably along the Potomac River. In 1820, for example, 62 percent of the enslaved population lived in Truro Parish. Truro also contained many of the largest and most famous slaveholdings—Mount Vernon, Woodlawn, and Ravensworth, to name a few. In 1820, 22 percent of slaves in Truro Parish lived on holdings with more than forty slaves, compared to only 6 percent in upper Fairfax Parish.[7]

A detailed map of Fairfax County that dates from the Civil War and was used by the Army of Northern Virginia also indicates a high concentration of farms and plantations near the river, especially within a few miles of Alexandria and along the Accotink Turnpike (which ran roughly parallel to the river toward Occoquan in Prince William County). It is unclear, however, how many of the farms were in fact slaveholdings, as the local slave population had diminished substantially by the Civil War and non-slaveholding farms are indistinguishable from slaveholding farms on the map. Several farms were also located near Bailey's Crossroads and Falls Church (both towns within easy reach of Alexandria and Washington), as well as along the northern banks of the Occoquan River (which serves as the southern boundary between Fairfax and Prince William Counties). The interior and western sections of Fairfax County—especially near Fairfax Court House, Centreville, and Dranesville—appear to have been less densely populated. In 1845 one Quaker spoke to a northern man who had bought a farm and settled with his family near Dranesville in western Fairfax, and asked him "what could have induced him to . . . settle here on an exhausted soil and in a neighborhood so sparsely inhabited that he must feel the want of society." The Quaker further claimed that "at a distance of 15 or 20 miles from the District of Columbia, the traveller finds himself in a wilderness of pines, and journeys for miles without seeing a single habitation." The general trend, thus, appears to have been a higher population density and larger slaveholdings near the Potomac River, especially near Alexandria and Washington, with inland and western Fairfax County being less densely populated and containing smaller slaveholdings.[8]

The sexual composition of the enslaved population in nineteenth-century Fairfax County appears to have been conducive to family formation, at least in theory. To gain an understanding of the local sex ratios between enslaved men and women of reproductive age (between fourteen and forty-five years old), it is useful to analyze the census data for the years 1820 and 1850. The reasons are twofold. First, the census returns for these two years allow for an easy calculation of sex ratios for the ages mentioned above, contrary to the census returns for other years such as 1810 or 1840, both of which list ages in different categories. Moreover, a comparison of data from 1820 and 1850 can reveal demographic developments over time, if there were any. The figures present a clear picture. In 1820 there were 2,062 slaves of reproductive age living in Fairfax County, of whom 1,073 were male and 989 were female—a relatively small difference of eighty-four, meaning that eighty-four males may not have been able to secure a spouse within the county. The male/female sex ratio was at that time 1.1. In 1850 the sexual composition of the local slave population was even more balanced. Of the

1,468 slaves of reproductive age living in the county at that time, 729 were male and 739 were female. The male/female sex ratio in 1850 was virtually balanced at 1.0, making family formation a theoretical possibility for almost all slaves.[9]

Building Momentum: Georgetown District, South Carolina

From the outset, exceptionally large slaveholdings dominated the fertile rice country of the Carolina lowcountry, which in the eyes of many contemporaries more closely resembled a Caribbean slave society than an American one. During the nineteenth century local slaveholdings only grew larger, and nowhere was this growth more pronounced than in Georgetown District. Indeed, throughout the antebellum period slaveholdings containing hundreds of bondspeople were far from uncommon in Georgetown District.[10]

Table 5.2 illustrates that as early as 1800, slaveholdings in the rice country provided a marked contrast with those in northern Virginia. In that year not only did 87 percent of the enslaved population live on plantations, but 44 percent lived on holdings with more than one hundred slaves. William Alston, the largest slaveholder at the turn of the century, owned no fewer than 550 bondspeople. By 1850, a year that marked the pinnacle of wealth and slaveholding in the district, local rice planters held 74 percent of the enslaved population on plantations with more than one hundred slaves, while 46 percent lived on holdings that contained more than two hundred. When South Carolina seceded from the Union, some sixty slaveholders in the district owned more than one hundred slaves, the largest of whom were the heirs of Joshua John Ward (1,131 slaves) and the former South Carolina governor Robert F. W. Allston (631 slaves). In Georgetown District, the six decades between 1800 and 1860 were thus characterized by significant expansion and consolidation in slaveholding size. The slave population itself increased by 53 percent, from 11,816 in 1800 to 18,034 in 1860. Both of these factors provide a stark contrast with northern Virginia, where slaveholding size and the slave population shrank over time.[11]

A number of factors contributed to the steady increase in slaveholding size during the first half of the nineteenth century. Most important, the rice industry had always favored economies of scale and consistently remained profitable enough to allow for expansion in land and slaves, in contrast to the grain industry in Fairfax County. As early as the eighteenth century, setting up a rice plantation in the Carolina lowcountry

Table 5.2 Distribution of the Georgetown District Slave Population by Slaveholding Size, 1800–1860

Year	Total Slaves	1–10 Slaves #	1–10 Slaves %	11–20 Slaves #	11–20 Slaves %	21–40 Slaves #	21–40 Slaves %
1800	11,816	614	5	937	8	866	7
1810	13,790	709	5	726	5	916	7
1820	15,547	648	4	651	4	997	6
1830	17,727	689	4	518	3	646	4
1840	16,018	576	4	437	3	790	5
1850	17,894	690	4	589	3	770	4
1860	18,034	826	5	674	4	545	3

Source: U.S. Census, 1800–1860, National Archives and Records Administration.

necessitated substantial capital. Large tracts of land needed to be acquired, and clearing, draining, and damming the swamps was costly, as was the acquisition of heavy machinery such as pounding machines. Numerous slaves—as many as thirty to forty from the outset—were needed not only to prepare the land for rice planting but also to produce large quantities of rice that would justify the heavy investments already made. Slaveholdings in the rice country were therefore already large to begin with, even in the eighteenth century.[12]

Nineteenth-century rice planters, most of them equipped with inherited wealth and all safeguarded by a booming rice industry, continued to make heavy reinvestments and produce rice on a large scale, expanding operations whenever possible. Some continued to clear new tracts of land and expand the rice industry to its geographic limits. J. Motte Alston, one of the few local planters who literally started from scratch in 1840 (with nothing but a 600-acre tract of swampland that he had inherited from his grandfather's estate), noted in his memoirs that high starting costs necessitated economies of scale. Clearing the swampland was "an expensive undertaking. These lands had to be levied, or banked in, to keep out the tide water." Dozens of slaves were needed; Alston started with a labor force of eighty enslaved men and women. Steam threshing mills, he recalled, cost roughly $8,000 before the Civil War, and new pounding mills could run as high as $20,000. Beginning planters who could not afford such mills had to pay their neighbors to thresh and pound their rice for them. However, the lure of immense profits convinced starting planters such as Alston that by producing large quantities of rice their plantations would soon begin to pay for themselves. Alston recalled nostalgically that a "tierce" (600 lbs.) of clean rice could be sold in Charleston for between $16.50 and $30, although "only

41–60 Slaves		61–80 Slaves		81–100 Slaves		101–200 Slaves		+200 Slaves	
#	%	#	%	#	%	#	%	#	%
1,523	13	1,235	11	1,431	12	3,294	28	1,916	16
1,281	9	914	7	1,286	9	3,826	28	4,132	30
1,292	8	1,163	8	998	6	4,463	29	5,335	34
926	5	1,309	7	1,729	10	6,343	36	5,567	31
610	4	677	4	1,735	11	5,297	33	5,896	37
511	3	963	5	1,168	7	4,986	28	8,217	46
668	4	1,125	6	1,651	9	5,659	31	6,886	38

the very choice 'head rice' brought the last named price." Moreover, it took "about eighteen or twenty bushels of rough rice, i.e., threshed rice, to yield one tierce of clean rice," making cultivation on a large scale a necessity. For those who inherited their plantations fully improved and with slaves, economies of scale were still necessary, not only to maintain operations and the lavish standard of living the owners enjoyed but also of course to increase their wealth even more.[13]

Indeed, most antebellum planters in Georgetown District—even those who started their careers in debt—amassed a great deal of wealth from their rice operations. According to Robert F. W. Allston, "the profits of a Rice plantation of good size and locality are about eight per cent per annum, independent of the privileges and perquisites of the plantation residence." Consider again the case of J. Motte Alston, who started from scratch in the winter of 1840–41, having to carve his Woodbourne plantation from a tract of overgrown swampland situated between the Waccamaw and Pee Dee rivers. Coming from a family of wealthy planters, his name was enough to secure the necessary credit from a Charleston rice factor to start operations. "Any amount of money I wanted in reason (from $1,000 to $20,000) I could have and no security asked," Alston later recalled. "At one time I was $45,000 in debt." However, rice cultivation proved so profitable in the decades preceding the Civil War that Alston was easily able to "return . . . the last cent, and before the opening of the war we [the rice planters] were all fairly rich men." By the time Alston was ready to retire in 1858—after just seventeen years as a rice planter—he had "accomplished what I had worked for, had paid off a heavy debt of $45,000 and had about $135,000 in what was then considered the very best investments in South Carolina." Moreover, having had no need to sell any of the eighty slaves he had bought when he began operations, his slave population had increased "from 100 to 150 as the years rolled by."[14]

The careers of other rice planters were even more spectacular. Robert F. W. Allston, whose father died young and thus failed to amass the fortunes of his neighbors, began in the early 1820s with an inheritance of only thirty-three slaves and a half share in a Pee Dee rice plantation on which, according to historian William Dusinberre, "everything remained to be done—clearing, ditching, draining—before the rich soil would be fit for cultivation." Starting relatively small, Allston cleared and brought into cultivation his 370-acre estate on the Pee Dee, purchasing large groups of slaves every few years as his production increased. As early as 1830 Allston was listed in the census schedules as the owner of 123 slaves; by 1840 this number had more than doubled to 264. Possessing excellent managerial skill and riding the waves of a booming rice industry, Allston was able to increase both his acreage and his slave labor force by the hundreds in the years that followed. By 1850, as he embarked upon his most successful decade as a rice planter—his 1853 rice crop alone sold for a whopping $43,500, yielding net profits of about $35,000—he owned 401 slaves (fifty-eight of whom he inherited from his aunt in 1840). When in 1859 he decided to buy 157 slaves (in two separate groups), his credit was good enough that he could issue his own personal notes. By 1860 Allston owned 631 men, women, and children in bondage, as well as an additional 119 he had bought to set up his son Benjamin in the planting business. In sum, Allston's successful rice operations afforded him the opportunity to increase his number of slaves from thirty-three in 1824 to 750 in 1860, only ninety-one of whom he acquired through inheritance. By that time Allston had moreover come into the possession of no less than seven plantations.[15]

Numerous other examples illustrate the relative ease with which rice planters in antebellum Georgetown District expanded their slaveholdings, both by purchasing more slaves and whenever possible by allowing their slave populations to increase naturally, despite high mortality rates. When Joshua John Ward began his career after his father's death in 1828, he had one large plantation to his name. In the 1830 census Ward is listed as the owner of 273 slaves, already a substantial number. Over the years, however, he "consistently reinvested his profits in more land and more slaves," according to historian George Rogers. By 1840 his slave labor force had more than doubled to 585, and during the next decade it peaked at over a thousand. By the time of his death in 1853, Ward possessed six plantations (Longwood, Springfield, Brookgreen, Prospect Hill, Alderly, and Oryzantia) and

1,092 slaves. A less spectacular—but still impressive—example is the career of Sampit River planter John Harleston Read I, who began operations at Maryville plantation in 1811. During the course of his long life as a rice planter, Read managed to double his number of slaves. The 1820 census lists Read as the owner of 104 bondspeople. In the next decade Read's slave population barely increased at all—in 1830 he still owned only 107 slaves, a testament to the high mortality rates in the lowcountry. By 1840 he owned 142 slaves, however, and in 1850 his slave population was up to 206. Along with his slave labor force Read also expanded his landholdings, purchasing the neighboring plantations of Oakley, Lucerne, and Upton. In short, rice planters in antebellum Georgetown District found numerous opportunities for expansion in land and slaves, unlike their counterparts in Fairfax County.[16]

While slaveholdings in Georgetown District were exceptionally large, most of the slaves owned by the largest slaveholders were in fact held on several separate—though often neighboring—units. One traveler to the district noted that it was common for "a gang of three or four hundred [slaves to be] divided and settled down upon different parts of the same estate, at a convenient distance from each other." Frederika Bremer, a Swedish visitor to the Poinsett plantations on the Pee Dee in the early 1850s, wrote that her host held his slaves on separate units that appeared "not to have more than sixty negroes upon them." Dr. Edward Thomas Heriot, the owner of 369 slaves, held most of his slaves on three separate plantations: 115 at Mount Arena, 125 at Northampton, 125 at Dirleton, and four domestic servants at his summer house.[17]

An examination of the Waccamaw Neck in lower All Saints Parish, a narrow peninsula between the Waccamaw River and the Atlantic Ocean, and the wealthiest section of the district, illustrates this trend. In 1860, the five southernmost plantations of the peninsula (Michaux, Calais, Strawberry Hill, Friendfield, and Marietta), plus four more located nearby (Youngville, Rose Hill, Forlorn Hope, and Clifton) were all owned by William Algernon Alston Jr. Alston held a total of 567 slaves, but they were spread out over nine units—an average of 63 slaves per unit—although the first five and the last three plantations were all adjoining, suggesting that the separate slave populations on Alston's plantations may have had relatively easy access to one another. Plowden Weston held a total of 334 slaves in 1860 on four adjoining plantations (Waterford, Hagley, Weehawka, and True Blue). The heirs of Joshua John Ward controlled 1,092 slaves on six separate plantations, located up and down the Waccamaw Neck in groups of two. The 335 slaves belonging to the estate of John Hyrne Tucker in 1860 were held on two adjoining plantations, Litchfield and Willbrook. Research by George Rogers on the Georgetown rice planters of 1850 indicates that the trend of multiple plantation ownership was also widespread in the other rice planting areas of the district—along the Pee Dee, Black, Sampit, and Santee Rivers. In sum, slaves owned by the largest slaveholders tended to be held in smaller—though compared to Fairfax by no means small—units than the census schedules tend to suggest. Moreover, most rice planters tried to consolidate their slaveholdings on adjoining plantations.[18]

The distances between slave villages of different plantations probably varied somewhat throughout the district. The rice plantations were all located along waterways, but the lower Waccamaw in particular appears to have been more densely populated than the other rivers. The Waccamaw Neck thus consistently contained not only the wealthiest plantations and the highest concentration of slaves but also the shortest distances between plantations, indicating a relatively close proximity between slave communities. In 1825 on this narrow peninsula, twenty-four miles long and nowhere more than four miles wide, some 3,970 slaves (26 percent of the total slave population) were living on thirty-five plantations. In 1860 there were 4,383 slaves (24 percent of the total slave population) on thirty-five plantations. Mills's Atlas of Georgetown District, surveyed in 1820 and improved in 1825, reveals that especially the southern tip of the peninsula contained a cluster of plantations, though many were of course the adjoining units of the same owner. Most of these plantations also claimed rice fields on the other side of the river, but the slave villages, like the rest of the plantation buildings and outbuildings, were probably located on the peninsula. Further north, the distances between plantations were greater; contacts between some of the northernmost plantations were even severed by numerous creeks and waterways. J. Motte Alston, who set up operations in the northernmost point of the Waccamaw Neck was separated from his neighbors' slaveholdings by Bull Creek. He recalled in his memoirs: "Cut off as we were from the outer world, there were no passers-by; those who came to Woodbourne came to see us. From Waccamaw Neck there was no approach. . . . Those who cared enough for us to make us visits, of course, had to be invited to remain to dinner or for the night." The Reverend Alexander Glennie, employed by many Waccamaw planters to preach to the slaves, could indeed only reach Woodbourne by river from neighboring Longwood plantation (owned by J. J. Ward), recording that "a boat conveyed us three miles up the river to Woodbourne."[19]

In other parts of the district the distances between plantations were greater and the slave populations likely more isolated. Mills's Atlas indicates that plantations along the Pee Dee, Black, Sampit, and North Santee rivers were more spread out, often a few miles apart. Moreover, the number and density of plantations decreased in direct proportion to their distance from Winyah Bay. On both sides of the Black River, which bisects the entire district, there were only about twenty-one plantations in 1825, and according to George Rogers, some twenty-eight plantations in 1850. None of these were located farther inland than the point at which the Black Mingo Creek joins the Black River, a bit more than halfway. Along the Pee Dee, which also traverses the entire district, there were approximately twenty plantations in 1825; twenty-one in 1850. These, too, were spaced farther apart as the distance from Winyah Bay increased and were moreover owned by only nine planting families. The Singleton plantation in the north appears to have been located at least eight miles from the nearest neighbor. Frederika Bremer described the Pee Dee plantation which she visited in 1853 as a "solitary, quiet abode; so solitary and quiet, that it almost astonishes me to find such an one in this lively, active part of the world, and among those company loving people." The memoirs of Elizabeth W. Allston Pringle, who grew up "next door" to the Poinsett plantation at which Bremer was a guest, confirm the Swedish traveler's observation. "[The] plantations were large," Pringle wrote, "so the neighbors were not near. . . . There were Mr. and Mrs. Poinsett at the White House, eight miles south of Chicora [Wood plantation]. . . . Mr. and Mrs. Julius Izard Pringle lived at Greenfield, eight miles southwest of us." The Sampit and North Santee rivers provide a similar picture. The trend appears to have been shorter distances between plantations near Winyah Bay, especially on the lower Waccamaw Neck, with fewer and more widely scattered plantations as the distance from the bay increased.[20]

The sexual composition of the enslaved population in Georgetown District remained relatively stable throughout the antebellum period. Census records indicate that in 1820 some 8,174 slaves (52 percent of the total slave population) were of reproductive age. Of this group, 4,081 were men and 4,093 were women—a negligible difference of only twelve. The ratio of males to females in that year was for all intents and purposes fully balanced at 1.0. The slave schedules for 1850 show a widening gap between men and women of reproductive age, though nothing approaching gross imbalance in the sexual composition of the local slave population. In 1850 there were 10,125 slaves of reproductive age (57 percent of the total slave population), of whom 4,956 were men and 5,169 were women—a difference of 213. The male/female ratio in that year was 0.96. That means that theoretically, over two hundred female slaves of reproductive age might have been unable to secure a partner within the county lines in 1850. Out of a total slave population of 17,894, however, this still constituted only a little over 1 percent of the population, but 4 percent of women of reproductive age. Such data indicate circumstances relatively conducive to slave family formation.[21]

Skewed Mushroom Growth: St. James Parish, Louisiana

The adoption of sugarcane as a staple crop at the turn of the nineteenth century triggered a plantation revolution throughout southern Louisiana, where a classic society with slaves was rapidly transformed into a firmly rooted slave society. The evolution of slaveholding size in St. James Parish clearly reflects the wider trend in the sugar country as a whole, as slaveholding size grew markedly during the antebellum period. However, contrary to the Upper South or the Carolina lowcountry, St. James Parish still resembled in many ways a developing slave society. The sugar industry was young, and many early sugar planters started with a moderate number of slaves compared to the grandees of Georgetown District, who often inherited established plantations with dozens or even hundreds of slaves. In St. James, most planters steadily expanded their operations either once their plantations began to pay for themselves or when they could secure the necessary extra credit. To be sure, local slaveholding size increased tremendously with each passing decade, but it never approached the grandeur of Georgetown District. Compared to Fairfax County, however, and indeed the rest of the South, the sugar plantations of St. James were large and successful.

Table 5.3. Distribution of the St. James Parish Slave Population by Slaveholding Size, 1810–1860

Year	Total Slaves	1–10 Slaves #	1–10 Slaves %	11–20 Slaves #	11–20 Slaves %	21–40 Slaves #	21–40 Slaves %
1810	1,755	678	39	397	23	144	8
1820	3,081	768	25	490	16	631	20
1830	5,024	949	19	624	12	878	17
1840	5,704	1,081	19	771	13	711	12
1850	7,724	1,505	19	860	11	916	12
1860	8,128	1,334	16	831	10	862	11

Source: U.S. Census, 1810–1860, National Archives and Records Administration.

Table 5.3 illustrates the steady growth of slaveholding size in the antebellum period. In 1810, most bondspeople in St. James Parish lived on small holdings containing twenty slaves or fewer (approximately 62 percent). Only fourteen slaveholdings constituted plantations, the largest of which was that of Jacques Roman, with seventy-nine. By 1820, however, the mushroom growth of local slaveholding size was clearly under way, with 59 percent of the local slave population living on plantations, three of which contained more than one hundred slaves. By the time the Civil War broke out, fully 40 percent of the local slave population lived on plantations containing more than one hundred slaves, a few of which contained over two hundred. The widow of Maryland-born Benjamin Winchester was the largest slaveholder in the parish, owning 240 bondspeople on her Buena Vista plantation. Next came the Lapice brothers, who jointly owned a plantation that contained 234 slaves. Valcour Aime owned 224 slaves. All in all, local slaveholding size increased exponentially during the antebellum period. The absolute number of bondspeople in the parish had likewise mushroomed by no less than 363 percent, from 1,755 in 1810 to 8,128 in 1860. Population growth was often erratic, however, as will become clear below.[22]

As in Georgetown District, the evolution of slaveholding size in St. James Parish can be explained by the fact that the sugar industry in Louisiana favored economies of scale and remained consistently profitable enough to allow for expansion in land and slaves. From the outset, starting and operation costs in the sugar country were extraordinarily high and only increased with time. Planters therefore required increasingly large numbers of slaves to produce enough sugar to justify the expenses. J. A. Leon claimed in his report on the Louisiana sugar industry that "slave-grown sugar, to be profitable to the planter, must be made on a large scale, and therefore requires a

41–60 Slaves		61–80 Slaves		81–100 Slaves		101–200 Slaves		+ 200 Slaves	
#	%	#	%	#	%	#	%	#	%
313	18	223	13	0	0	0	0	0	0
397	13	278	9	169	6	348	11	0	0
293	6	342	7	629	13	1,309	26	0	0
547	10	259	5	448	8	1,424	25	463	8
506	7	829	11	445	6	1,397	19	1,176	15
829	10	696	9	363	4	2,515	31	698	9

large capital.... Money must be procured to buy negroes, land, provisions, materials for extensive buildings, [and] machinery." According to one estimate, the capital investment for a new sugar plantation in 1800—based on 100 arpents of land and forty slaves—averaged $61,000. In the decades that followed, however, and especially after 1830, capital investments increased tremendously. Between 1836 and 1840, Samuel Fagot reportedly spent almost $200,000 to set up operations on his Constancia plantation, including the acquisition of a labor force consisting of 130 slaves. Frederick Law Olmsted was told by a Louisiana sugar planter in 1856 that no less than $150,000 should be invested in starting a new sugar plantation. As a result, a vast majority of aspiring planters began their careers "very heavily in debt."[23]

Acquiring land and a slave labor force already required a significant amount of capital, but as Leon noted, Louisiana planters had to additionally invest in the high-tech machinery necessary to process cane into granulated sugar and molasses. The precarious nature of sugar manufacturing, especially the crop's vulnerability to both frosts and rapid spoilage, required virtually all sugar planters to invest in a private sugar mill instead of having their cane processed at a neighbor's mill (as was often the case with rice in Georgetown District). In the opening decades of the nineteenth century, a primitive but effective vertical mill, consisting of three wooden rollers and driven by animal power, could be built from local materials, costing the planter between $1,000 and $2,000. But after the first steam engines were employed in 1822, the costs for constructing and equipping a sugarhouse truly skyrocketed, as did the number of slaves planters needed to ensure a return on their investments. Traveler Joseph Holt Ingraham noted in 1835 that "it requires almost a fortune to construct [a sugar mill]." Another visitor named James Stuart, who stayed at a sugar plantation in 1833, was told by his host that the "sugar-mill on this property cost about 12,000 dollars." Between 1830 and 1860, numerous other technological improvements, such as the introduction of vacuum pans, steam evaporators, and even refinery equipment, made the investment in machinery even more expensive. A visitor to St. James Parish in the 1850s claimed that "on every plantation the sugar house is one of the most prominent objects. It would be impossible to give a correct idea of the immense amount of money lavished upon these adjuncts to the sugar estate." Local planter J. B. Armand spent $51,000 in the 1840s on a "complete double effect apparatus, composed principally of four cylindrical evaporators." The growing number of jointly owned and operated sugar plantations in St. James Parish during the last three decades of the antebellum period testifies to the exorbitant costs in acquiring the most advanced sugar-producing machinery.[24]

A perusal of the agricultural census for 1860 indicates that on the eve of the Civil War, the average value of sugar-processing equipment alone on the largest plantations in St. James Parish (those containing fifty or more slaves) was $21,220, but on some plantations it was as high as $55,000. Significantly, the plantations containing the most expensive equipment in 1860 also contained the largest numbers of slaves. Constancia plantation, which contained 131 slaves, had $55,000 worth of sugar-making machinery in 1860. Golden Grove, which contained 187 slaves and was the fourth largest plantation in 1860, had the distinction of being the only plantation in the parish with two sugarhouses. The three slaveholdings in the parish that contained more than two hundred slaves (those of Madame Winchester, the Lapice brothers, and Aime), all boasted expensive new refineries.[25]

Despite exorbitant setup and operation costs, however, it was not uncommon for some aspiring planters to start small and gradually expand both their land and slaveholdings over a period of several years. Historian J. Carlyle Sitterson, in his classic study of the antebellum sugar industry, found that "although the large farm or plantation became the dominant agricultural unit in the southern cane region, small and medium sized cane farms remained significant even until the close of the antebellum period." Sitterson added that even after the invention and widespread adoption of steam-powered sugar mills in the 1820s and 1830s, small cane farmers often "clung tenaciously" to the old-fashioned animal-powered mills, which may have been less efficient but required less capital outlay and thus fewer slaves. Historian Mark Schmitz also argued that in antebellum Louisiana, "not all of the horse power farms were so unsuccessful. Many of them merged, switched to steam, or survived with their basic apparatus." And John C. Rodrigue, in his study of the transition from slavery to freedom in the Louisiana sugar country, likewise found that "it was not uncommon for an enterprising individual to start small and with a little luck grow two or three successful sugar crops that would propel him into the budding upper echelon."[26]

The number of small cane farms decreased over time, but even as late as the 1850s many sugar-producing estates in St. James Parish held significantly fewer slaves than one might expect, often between eleven and twenty. In 1850, for example, Benjamin Schexnaydre produced on his 115-acre estate some fifty-five hogsheads of cane sugar and 3,200 gallons of molasses with only twelve slaves. Schexnaydre's sugar-making equipment was valued at

$1,000, a very low capital investment indeed, indicating that he almost certainly processed his sugar in an old-fashioned animal-driven mill. (Significantly, eleven of Schexnaydre's slaves were adults. Therefore, the number of working hands on his smallholding was much larger—and the value of his slaves much higher—than it would have been on a farm of similar size in Fairfax County, where most smallholdings also contained children and thus fewer working hands.) Schexnaydre disappears from the slave schedules in 1860, perhaps because of death, emigration, or bankruptcy. Many smallholdings in the sugar country were volatile and susceptible to failure. As Ira Berlin has pointed out, "for every aspiring master who climbed into the planter class, dozens failed because of undercapitalization, unproductive land, insect infestation, bad weather, or sheer incompetence." Whether this was the case with Benjamin Schexnaydre or not, numerous local smallholders come and go from the antebellum census schedules, suggesting that not all were able or willing to keep their businesses afloat.[27]

The career of Andrew Crane provides an example of a relatively successful late-antebellum sugar planter who began with a limited number of slaves and gradually expanded his holdings, despite numerous setbacks. Starting operations in 1849 on a narrow 225-acre tract of land, located on the east bank of the Mississippi, Crane had not yet produced his first cane crop by 1850. He owned only nine slaves, consisting of four adult men, three adult women, and one twelve-year-old boy. In the years to come, however, he purchased more slaves and began to cultivate cane commercially. In 1850 alone, Crane purchased eight slaves at four different auctions in New Orleans, including one group of four that cost him $3,325. Within a year, two of his slaves attempted to run away, burdening him not only with the loss of their labor but also the costs for newspaper advertisements and jail fees. Apparently running into financial difficulties with his creditors, Crane lost three slave men in April 1852, who were seized by J. Corning and Co. and sold at public auction. Still, he managed to stay in business. The 1860 census enumerator found Crane just barely slipping into the ranks of the planter class as the owner of twenty-one slaves, most of whom were adult males. His sugar-making machinery was valued at $12,000, and his production in 1860 consisted of 100 hogsheads of cane sugar in addition to 7,000 gallons of molasses. Although most slaveholders in St. James Parish were managing large-scale operations by the end of the antebellum period, a number of ambitious cane "farmers" such as Schexnaydre and Crane kept average slaveholding size lower than it was in Georgetown District.[28]

Like the Carolina lowcountry, but unlike northern Virginia, slave-based agriculture in southern Louisiana generally provided numerous opportunities for expansion in slaves. The profits procured from the booming sugar industry allowed the more successful smallholders such as Crane the opportunity to climb a notch higher in the slaveholding hierarchy, while those who entered the business with inherited wealth or in joint partnerships, or had made their fortunes elsewhere, were often able to increase their already large numbers of slaves into the hundreds. As early as 1800, according to Sitterson, a sugar planter could expect at least a 14.6 percent return per year on his invested capital, allowing him to retire his debts within a short period of time, reinvest in more slaves, and become considerably wealthy. Early visitors to the region frequently exclaimed that "those who have attempted the cultivation of the Sugar Cane are making immense fortunes." Estwick Evans, who visited the sugar country in 1818, claimed that "the planters here derive immense profits from the cultivation of their estates. The yearly income from them is from 20,000 to 30,000 dollars." When the Duparc brothers of St. James took over the family plantation in the 1840s, it "had seen 20 straight years of healthy profits." By the end of the antebellum period, the largest sugar planters had become grandees, commanding small armies of slaves and living aristocratic lives on estates valued in the hundreds of thousands of dollars. T. B. Thorpe, who visited one of the largest plantations in St. James Parish in 1852, reported that his host had made a profit of $79,600 that year, after deducting all annual expenses. Some 215 slaves lived on the plantation, and the estate was valued at $700,000.[29]

Valcour Aime's plantation (nicknamed "Le Petit Versailles") provides a good example of the profits and expansion in slaves of which local sugar plantations were capable when managed wisely. Aime began his career as a sugar planter in 1820 with substantial inherited wealth, which allowed him to start operations on a large scale. As early as 1816, when he married the daughter of Jacques Roman—one of the largest slaveholders in the parish—his assets already totaled $75,000. Together with his wife Josephine's dowry, the couple started out with over $94,600. In 1820, Aime bought a plantation located on the west bank of the Mississippi for the sum of $170,000. The 1820 census lists him as the owner of forty-eight bondspeople. By 1823 Aime was already producing a respectable cane crop; his 112 hogsheads of sugar sold that year for approximately $6,160. By 1830, Aime had applied steam power to his sugarhouse, was producing 395 hogsheads of sugar per year, and owned 103 slaves: fifty-three adult males, thirty-five adult females, and fifteen children. In the decade that followed, Aime again practically doubled his slave population, which by 1840 consisted of

201 bondspeople. That year he produced 918 hogsheads of sugar. With a keen eye for increasing efficiency, Aime spent over $10,000 on new sugar-making equipment in 1843, and in 1845 he even traveled to Cuba to inspect Caribbean methods of refinement, which he subsequently introduced on his own plantation in St. James. By the time the Civil War broke out in 1861 Aime was producing over a thousand hogsheads of sugar a year and was the master of 224 slaves. His wealth and lavish lifestyle were legendary throughout Louisiana.[30]

Several of Aime's peers also consistently increased their slaveholdings during their relatively short careers. Jacques Roman, one of the earliest and most successful sugar planters in the parish, already owned seventy-nine slaves in 1810. By 1820 his estate held 112 slaves, then controlled by his widow, and by 1830 Roman's slave population had increased to 165. J. L. Fabre, another early planter in St. James Parish, owned seventy-six slaves in 1810. Within a decade his slave population had increased to 102, and during the decade that followed it peaked at 147. Benjamin Winchester held eighty-two slaves on his Buena Vista plantation in 1830, of whom twenty-six were children under the age of ten. By 1840 he had increased his number of slaves to 133, and in 1850 he was listed in the census schedules as the master of 196. By the eve of the Civil War, Winchester's widow held 240 men, women, and children in bondage, almost triple what her late husband had owned thirty years earlier. Former Louisiana governor A. B. Roman increased his number of slaves from 118 in 1850 to 142 in 1860. The list goes on and on.[31]

Not all slaveholdings grew with such consistency, however, even on the largest plantations. For some, short spurts of population growth were followed by stagnation or even decrease, and vice versa. Samuel Fagot, for example, began operations on his Constancia plantation with some 130 slaves in 1840. By 1850 this number had increased to 156, but during the decade preceding the Civil War, Fagot's slave population shrank back to 131 (at least six slaves absconded in 1855). Golden Grove plantation, owned by the Shepherd brothers, contained 262 slaves in 1840. By 1850 their slave population had decreased slightly to 249, but a decade later it had shrunk considerably to 187. E. J. Forstall held 129 bondspeople in 1850; by 1860 this number had been reduced to 103. Other slaveholdings only increased slightly or stayed relatively the same. The slave population of George Mather's widow, for example, only barely increased between 1850 and 1860, from 133 to 139. Even Aime's slave population failed to grow between 1850 and 1860, when it decreased slightly from 231 to 224. What does this tell us about the evolution of slaveholdings in St. James Parish, and how can such cases be explained?[32]

In the broadest sense, cases of population decrease and low population increase indicate that the slave population of St. James Parish often failed to reproduce itself naturally. Many local sugar planters were therefore primarily dependent on the expensive domestic slave trade to expand their slaveholdings, which again points to the immense profits to be made in the sugar industry. Michael Tadman has argued that slave populations on sugar plantations throughout the Americas—including Louisiana—consistently experienced natural demographic decline. "The demographic experience of Louisiana's sugar slaves was unique by U.S. standards," according to Tadman. Despite substantial increase in slaveholding size, "the sugar parishes stood out as a miserable island of natural decrease in the midst of the otherwise consistent pattern of natural increase that stretched across the cotton, tobacco, and rice plantations of the United States." This was largely due to a combination of factors, including a gross sexual imbalance among the slave population (which will be further discussed below), as well as high mortality and low fertility rates, all of which resulted in a shortage of children on most plantations, as discussed in chapter 3. It was a phenomenon that did not go unnoticed by visitors to the region. Thomas Hamilton reported in 1835 that local sugar planters "must buy to keep up their [slave] stock, and this supply principally comes from Maryland, Virginia, and North Carolina."[33]

In St. James Parish in 1850, slave deaths exceeded slave births by 28 percent. In Fairfax County, by contrast, births exceeded deaths by 61 percent. The rice plantations had disastrous fertility and child mortality rates, but even in Georgetown District births exceeded deaths by 9 percent, indicating low population growth instead of natural decrease. Moreover, Georgetown District at least fit the profile of a slave society with high mortality rates. The slave population there increased by only 53 percent during the entire antebellum period, including imports. Between 1830 and 1860, it barely increased at all. In St. James Parish, however, the slave population increased by 363 percent during the nineteenth century, whereas naturally it should have decreased substantially.[34]

Slave population growth in St. James was moreover highly erratic in nature. Between 1810 and 1820, for example, the slave population grew by an extremely high 76 percent, suggesting that starting planters bought heavily at the slave auctions of New Orleans. In the following decade it grew again by 63 percent, twice the national average. In the 1830s there appears to have been a temporary lull in the slave-buying frenzy, when the population increased by only 13.5 percent (the national average was 23.8 percent). This may have been partly due to a series of restrictive laws passed between 1831 and 1834, which in the wake of the Nat Turner insurrection temporarily prohibited slave imports by professional traders. Slaveholding size, however, still increased significantly, suggesting that large planters turned to buying out their neighbors until the ban was lifted. In the 1840s, imports picked up again and the slave population jumped by 35 percent. In the decade preceding the Civil War, however, the population again stagnated, growing by only 5 percent (compared to the national average of 23.4). Although when one takes into account that in 1850 slave deaths exceeded births, even a population increase of 5 percent means that in 1860 an estimated 10 percent of the slave population (around 834 slaves) had probably been recently imported from outside the parish.[35]

As a result of massive slave imports, triggered by the efficiency of economies of scale in sugar production and financed by the spectacular profits the industry generated, enslaved people in St. James Parish increasingly lived on large holdings by the late antebellum period. The prevalence of large slaveholdings did not, however, mean that slaves found themselves isolated on enormous estates. Indeed, perhaps nowhere else in the South were tracts of land so narrow and plantations located so close to each other.

The original land grants by both French and Spanish colonial governors had been marked out in such a way as to assure settlers frontage on the Mississippi River, which bisects the parish and divides it into east and west banks. Tracts therefore narrowly fronted the river and extended back at right angles. Single grants measured forty arpents in depth; double grants measured eighty arpents. In practice, planters in the late antebellum period increased their acreage by clearing the cypress forests behind their estates and bringing that land under cultivation as well. Some bought out their neighbors' landholdings to increase the river frontage of their estates, but in general plantations in St. James Parish remained almost comically narrow throughout the nineteenth century. The close proximity of sugar plantations throughout southern Louisiana was frequently commented upon by passersby. Victor Tixier, a visitor to St. James Parish, was "struck by the narrowness of the cultivated fields." Harriet Martineau, traveling north by river from New Orleans, wrote: "All the morning we were passing plantations, and there were houses along both banks at short intervals." Eleanor Parke Custis Lewis, the wife of Fairfax County slaveholder Lawrence Lewis, remarked on a similar visit to southern Louisiana that "the Houses are so near each other on the river that it looks like a village on each side." And Indiana agriculturalist Solon Robinson wrote during a visit to St. James in 1848 that "all along the [river] road, the small Creole places are thick as 'three in a bed.' . . . Fancy a farm three rods wide, and 480 rods deep, and if you like here is a lot on 'em."[36]

Several contemporary sources more precisely illustrate just how close these plantations were to each other, especially Champomier's *Statement of the Sugar Crop*—which mentions the precise distance from New Orleans (by river) of each plantation in the parish—but also Persac's meticulously surveyed 1858 map of the Mississippi River plantations. In 1846, Champomier listed twenty-eight plantations in twenty-one miles on the west bank. On the east bank there were thirty-nine plantations in twenty-two miles. Several mile markers counted multiple plantations. At mile sixty-six on the east bank, for example, were three sugar plantations, listed as belonging to "Noel Jourdan & Gaudin," "Ed. Jacob & Co.," and "P. & O. Colomb." Ben Winchester's plantation on the west bank, one of the largest slaveholdings in the parish, contained only twelve arpents (about ten acres) of river frontage and was one of three plantations located at mile seventy. Sósthene Roman's plantation, another large slaveholding containing ninety-two slaves in 1860, was advertised at the end of the Civil War as containing eighteen arpents of river frontage (fifteen acres). Persac's map illustrates several tracts of land that appear to be little more than slivers. These, too, were often sugar plantations. Arnaud LeBourgeois's fully equipped sugar plantation, which held twenty-four slaves in 1860, contained only four arpents (3.5 acres) of river frontage; the plantation of Francois Brignac was advertised as measuring a mere 2.25 by eighty arpents (two by sixty-eight acres). In short, plantations in St. James Parish were narrow, indicating that slave villages between separate plantations were probably located literally within sight of one another.[37]

As stated above, sugar planters' regular dependence on the domestic slave trade to expand their slaveholdings had much to do with a severe sexual imbalance on most plantations, which limited natural population growth among their slaves. As numerous scholars have pointed out, sugar planters throughout southern Louisiana—and indeed the Americas—preferred slave men to women. Tadman found that on most plantations, "the demographic problem stemmed from the priorities of the sugar planter. Sugar planters, unlike the great majority of [slave] owners, calculated that they could maximize profits by continually skewing their labor force toward men." According to Ira Berlin, two-thirds of the slaves bought by Louisiana sugar planters were men, and Richard Follett has calculated that 70 percent of slaves exported to New Orleans were men. Planters in St. James Parish followed the general trend, as is evident from the distorted sex ratios for the parish. Whereas in antebellum Fairfax County and Georgetown District the number of men and women of reproductive age remained roughly equal, in St. James Parish it never even came close. In 1820, some 1,858 slaves were between the ages of fourteen and forty-five. Of these, no fewer than 1,101 (59 percent) were men, while 757 (41 percent) were women. The male/female ratio for that year was 1.45. In practice, this meant that for 31 percent of the male slave population (344 slave men) there were no suitable marriage partners within the parish. By 1850, the situation had even worsened slightly. In that year the census enumerator listed 4,324 slaves of reproductive age, 2,619 of whom were men (61 percent) and 1,705 of whom were women (39 percent). That left a male/female ratio of 1.54 and an exceptionally high surplus of 914 men (35 percent of the male population). On average, then, marriage was probably possible for only two-thirds of enslaved men in St. James Parish, a situation that did not improve with time.[38]

Conclusion

In different communities of the nineteenth-century South, the spatial distribution and sexual composition of slave populations reflected the nature of regional agriculture. The limited number of bondspeople that grain producers in Fairfax County needed or could afford in the antebellum period triggered a significant decline in slaveholding size, as droves of surplus slaves were sold south via the domestic slave trade and plantations were reduced to small farms. In Georgetown District, where rice production favored economies of scale and consistently remained profitable enough to allow for reinvestment in more slaves, the size of local slaveholdings grew during the nineteenth century. By the time the Civil War broke out, most slaves in Georgetown lived on plantations that contained more than one hundred slaves, and many lived in relative isolation on enormous estates. In St. James Parish, where local agriculture also favored economies of scale and remained extremely profitable during the antebellum period, slaveholding size likewise grew tremendously, mostly as a result of slave imports from other southern states. Plantations increasingly contained numerous slaves and were located very close to each other. As a developing slave society, however, slaveholdings in St. James on the eve of the Civil War had not yet evolved to be as large as slaveholdings in Georgetown District. The sexual composition of slave populations in Fairfax County and Georgetown District were balanced, but in St. James Parish sugar planters imported far too many men, resulting in a severe sexual imbalance.

The social landscapes in all three regions greatly affected the boundaries and opportunities with which slaves were confronted when seeking a mate and had far-reaching consequences for marriage strategies and the threat of forced separation across time and space.

6

Marriage Strategies and Family Formation

A great deal of the historical disagreement concerning slave family life has revolved, and continues to revolve, around the issues of marriage strategies and family formation. For much of the twentieth century, historians of southern slavery believed that family formation was essentially nonexistent in the slave quarters, with interpretations varying from perceived character deficiencies among African Americans to the oppressive nature of bondage. Traditional views of slave families held that relationships between enslaved men and women had been fleeting and casual affairs, resulting in a preponderance of single mothers and unattached men. By consequence, slave families had been predominantly matrifocal and only loosely organized.[1]

In the 1970s, however, new evidence that long-term slave marriages and two-parent households were the norm throughout the South led scholars to reject the traditional paradigm, but different causal theories for the newly perceived stability of slave families added fuel to the fire of the emerging slave agency debate. Herbert Gutman, for example, argued that long-lasting marriages and two-parent households among slaves were largely the result of their "inner strength" and determination to assert their humanity in the face of a brutal and dehumanizing system. Fogel and Engerman, on the other hand, attributed the perceived prevalence of two-parent households among the slave population to the interference and incentives of slaveholders, whose interest in encouraging the establishment of cohesive slave families stemmed from their desire to discourage flight and increase the labor force through natural reproduction. More recent studies by historians such as Brenda Stevenson, Larry Hudson, Wilma Dunaway, and Emily West have complicated these views by consistently arguing that although marriage and family formation were indeed the norm among slaves, two-parent households were nowhere near universal. As yet the extent to which different domestic arrangements were adopted by enslaved couples, as well as the reasons for such arrangements, is still far from clear.[2]

In the very simplest terms, slaves could pursue one of two strategies when they sought to marry and establish a nuclear family. First, they could choose a spouse from the home plantation, in which case they formed co-residential marriages and two-parent households. Second, they could marry a slave (or free black) who lived outside of the home plantation, forming cross-plantation marriages and divided households (frequently called "broad" or "abroad" families). Certainly a number of factors influenced slaves' marriage choices—including, of course, love and physical attraction, as well as coercion by white masters—but as this chapter will show, the different social landscapes in various localities appear to have been at the root of slaves' decisions, as the nature of slaveholding in different regions confronted enslaved men and women with different boundaries and opportunities for marriage and family formation across time and space.[3]

This chapter examines the nature of slave family formation as determined by slaveholding size and sex ratios. How did enslaved people adapt to the physical and demographic boundaries of their containment when seeking a mate? What were the consequences for their domestic arrangements and family structures? By comparing families in different parts of the non-cotton South, this chapter will show that antebellum slaves usually strove to create two-parent households whenever possible; however, not all were able to realize that ideal, and those who could not adapted their marriage strategies and family lives accordingly.

Transcending the Geography of Containment: Fairfax County, Virginia

Ascertaining slave family structure in northern Virginia or any other region of the antebellum South provides a considerable challenge. Because slave marriages were not legally recognized, few nineteenth-century sources document family ties at all. In official sources only the mothers of small children are mentioned with anything approaching regularity (especially if they were to be sold together), but fathers are only very rarely indicated. Moreover, slave marriages were frequently ruptured, most notably by sale and estate divisions, but also by death and sometimes by voluntary separation. Elijah Fletcher, for one, observed on Thomson Mason's plantation in 1810 that sometimes "a man and woman will make some agreement between themselves . . . till they disagree, and then will part." Certainly not all slave marriages ended so abruptly, but the breakup of slave unions, whether forced or voluntary, meant that families' domestic arrangements were continually being established and reestablished over time, making it tricky for historians to reconstruct family ties.[4]

Despite these limitations, however, it is possible to provide a qualitative indication of how widespread certain types of domestic arrangements were by combining data concerning slaveholding size and sex ratios with the limited sources that are available. Chapter 5 concluded that slaves in Fairfax County consistently lived on relatively small holdings in the nineteenth century, while sex ratios among men and women of reproductive age remained roughly equal, theoretically making marriage and family formation possible for most. Under such circumstances, two trends might be expected: first, that cross-plantation marriages and divided households were especially widespread among slaves in Fairfax County (due to a lack of suitable partners on the home farm or plantation); and second, that the proportion of divided households increased with time (in direct correlation to the decline in local slaveholding size). An examination of the available evidence suggests that both hypotheses are valid.

The slave inventories of farms, where the majority of Fairfax County slaves lived throughout the antebellum period, provide plenty of indirect evidence of divided households by listing what on the surface appear to be single mothers and single men.[5] The 1826 inventory of George Triplett's estate lists ten slaves, all living in female-headed households: Sinah and three children, Betty and three children, Mary, and Eliza. Another farmer named Charles Guy Broadwater owned two slave mothers with a total of nine children, but the only male living on the estate was elderly. Numerous other examples of female-headed households can be found in the will books, and by the 1850s they appear with even more frequency.[6] The same goes for inventories that indirectly suggest cross-plantation relationships by listing single men who obviously lacked suitable marriage prospects on the home farm. Charles Smith owned ten slaves upon his death, consisting of one elderly man, four men of working age, and one woman with four children. It is unclear whether the woman was married to one of the adult men on the farm (they may have been her brothers), but for at least three of these men there were certainly no potential partners living on the farm. F. L. Lee's estate likewise listed six adult men but only two adult women with a total of four young sons. And in 1826 Sabret Scott was the owner of three adult men and one elderly woman named Milly. For enslaved men living on these and other small estates, cross-plantation relationships were their only options if they wanted to marry and establish a family.[7]

Only a handful of estate accounts from small slaveholdings list co-residential family units, which suggests that a few slaveholders may have either taken care not to split up families during estate divisions or sales, or perhaps even bought their slaves' spouses in an effort to keep families together. Charles Henderson's slaves, for example, consisted of a complete family: Lewis and Penny, and their children named Lucy and John. Mastrom Cockerill owned "Negro John & Lizza, his wife," along with their two children named John and Sophia. Many such co-residential slave couples most likely started out as broad couples but were fortunate enough to have been joined through purchase. Inventories that clearly list co-residential families are heavily outweighed, however, by inventories that indicate cross-plantation marriages and divided households. Moreover, it is important to remember that as adults the children of these co-residential couples would have lacked potential partners on the home farm, thereby perpetuating cross-plantation family structure among future generations of slaves. The ten slaves who belonged to the estate of James Burke in 1826 provide a case in point. Burke's slaves consisted of one elderly couple (Charles and Patty, both valued at nothing), their adult daughter Patty with her five children, and another adult daughter named Charlotte with her infant child. Both Patty and Charlotte grew up in a household with co-residential parents, but lacking any suitable partners on the farm, both women were forced to marry abroad as adults.[8]

The prevalence of single mothers and single men in the inventories suggests that cross-plantation marriages and divided households were already widespread among Fairfax County slaves during the early decades of the nineteenth century and that they increased with time. Yet it is important to consider that not all single adults in the region were married at any given time. The partners of some, for example, may have been deceased, sold, or otherwise removed from the area, but limited evidence also suggests that not all slaves desired permanent relationships, including some single mothers. Northern reporter James Redpath spoke with one young woman in Fairfax who had two small children, but when asked where her husband lived replied "I'se not married." Elijah Fletcher, the Vermont tutor at Hollin Hall, lamented the "promiscuous intercourse between the sexes and no regard to decency or chastity" among the slaves, which often resulted in the women having "as many children as a bitch will puppies." Despite the exaggerations of white observers, cases of casual sexual relationships were certainly not unknown in the slave South, yet they do not appear to have been as widespread as many contemporaries believed. Fletcher himself admitted that most slaves in Fairfax County "pretend to have wives and husbands, but have no ceremony in marriage."[9]

Indeed, one indicator that most cross-plantation relationships—certainly those that resulted in children—were not casual or fleeting is that enslaved people consistently availed themselves of opportunities to visit their spouses and family members whenever possible, thereby transcending what Stephanie Camp has called the "geography of containment" which bound them in space. Slaves' physical mobility was thus an important factor in family formation and family life in northern Virginia. While the law did not recognize slave marriages or grant slave couples the rights of married people, local slaveholders implicitly accepted their slaves' broad relationships and family connections by granting them permission to visit with friends and family members across property lines.[10]

As in the rest of the South, local slaveholders issued written passes—which legalized slaves' physical mobility between specific locales—for weekend visiting, usually between Saturday evenings and Monday mornings, pending good behavior. John and Orvill, two young slave men owned by David Wilson Scott in 1820, regularly requested and received passes on Saturday evenings to leave the plantation and "see their friends" on a neighboring plantation, probably their girlfriends. Sometimes passes were only issued on Sundays. English traveler John Davis recorded during his visit to Pohoke plantation in the early 1800s that only "on the Sabbath [are] the negroes at liberty to visit their neighbours." Outside of these designated times, and without passes, leaving the plantation or farm was strictly prohibited. The pass system was of course far from an attractive arrangement for divided couples. They were issued only for specific farms or residences and were valid for only limited amounts of time. As Stephanie Camp put it: "Passes gave some bondpeople permission to go to some places some of the time. Passes also prevented most enslaved people from going to most places most of the time." For those with few other options, however, passes offered the only opportunities to visit with spouses or children.[11]

The physical movement of slaves across space usually consisted of men traveling on Saturday evenings or Sundays to see their wives and children, or girlfriends, at their places of residence. That men often went to great lengths to visit their girlfriends and families contradicts arguments that they played only an indifferent role in family life when they courted and married across plantation lines.[12] Indeed, many slave men tried to stay with their wives longer than their passes allowed. Some hid out in the neighborhood of their wives' residence and remained truant for a period of time, while others attempted to escape the South altogether with their wives. Edward Dulin, the owner of Clover Hill, advertised the flight of his "Negro Emmanuel" in 1809, emphasizing that Emmanuel "has a wife living in the neighborhood from whom he has removed his clothes [and who has now] also absconded." Another truant named Andrew had "a wife at Mrs. Fendall's farm in Fairfax County . . . where it is probable he may be harbored by her at night, and skulk about the neighboring woods through the day." Townshend Derricks, who ran away in 1857, "took in his company with him his wife Mary, the property of the estate of Caldwell Carr." A northern abolitionist society received Derricks as a refugee, but reported that "the wife was captured and carried back," adding that it "was particularly with a view of saving [her] that Townshend was induced to peril his life, for she (the wife) was not owned by the same party who owned Townshend, and was on the eve of being taken by her owners some fifty miles distant into the country, where the chances for intercourse between husband and wife would no longer be favorable."[13]

Slaves and slaveholders also left ample evidence of the active role that enslaved men played in maintaining family contact across physical boundaries. George Jackson, a northern Virginia ex-slave who was interviewed by the Federal Writers' Project in the 1930s, claimed that when his "father wanted to cum home he had to get a permit from his massa. He would only come home on Saturday. He worked on the plantation 'joinin' us. All us chillun and my mother belonged to Massa Humphries." One slave from western Fairfax County, upon being asked by James Redpath whether he was "a married man," replied that he was "gwane to see my wife now." Redpath added: "He told me she lived some five or six miles off."[14] Anne Frobel, the daughter of a local slaveholding family that owned nineteen bondspeople in 1860, recorded in her diary at the start of the Civil War that "Cesar, one of our negro men, has a wife living some three or four miles away where he goes every Saturday evening." Charles, another slave who lived on the same farm, went to see his wife in Washington one Saturday evening and never returned. The diary of Richard Marshall Scott Jr., of Bush Hill, also mentions several instances of cross-plantation visiting and broad marriages among his slaves. On 10 July 1849, for example, Scott rode his horse "up to Ravensworth to see my man Joe, who is sick there at his wife's."[15]

A few of Scott's slaves were in fact married to free blacks, an arrangement that often carried certain advantages with it for both parties. A free black man married to a slave woman was spared the burden of having to care for her financially (although his children were legally slaves). The slave woman, on the other hand, might hope to eventually be purchased and freed, or at least annually hired, by a free black husband. (One free black man from Fairfax County named Griffin Dobson purchased his wife and five children in 1852 for the sum of $1,200, having gone all the way to California to earn enough money to unite his family.) A slave man married to a free black woman, which was less common, would have had free children. At Bush Hill there is evidence of both types of cross-status marriage, even in the early decades of the nineteenth century. In 1824 Richard Marshall Scott, Sr., "attended the funeral of negro Betty, a free woman, wife of my servant Moses." In 1846, a free black man named Thomas Grey, whose wife was owned by Scott (Jr.), "came out to work for me by the month in order to pay the hire of his wife." Ellen Ann, another slave on the farm, was the "wife of David Grey, free negro living near Claremont, to whom she is hired."[16]

As stated earlier, slaveholding size was not the only factor that influenced slaves' marriage choices. Love and physical attraction between slaves certainly did not always respect the fences that contained them, and some enslaved women—even those who lived on farms with a suitable number of prospective partners—purposefully chose husbands who lived on other holdings. With the exception of those married to free men, however, this was most likely a situation they felt compelled to tolerate rather than an ideal arrangement. Contrary to the assertions of some historians, slave women do not appear to have procured many "benefits" when they married across plantation lines. If anything, the opposite was true. Although husbands in northern Virginia often went to considerable lengths to visit their wives and children who lived on other slaveholdings, their absence from the family household during the week not only disproportionately burdened slave women with domestic and child-rearing duties, as argued in chapters 3 and 4, but it also made them vulnerable to the sexual advances of their masters and other white men. Incidents of rape were certainly not unusual in northern Virginia. One local slave claimed that "dey was many a little white-skinned nigger baby" where he grew up. The high proportion of mulatto slaves living Fairfax County during the last decade of the antebellum period painfully illustrates one of the possible consequences of cross-plantation family structure in the area. In 1850 some 22 percent of the slave population there consisted of mulattoes; by 1860 it was 30 percent. Although certainly not all mulatto slaves in the county were the direct products of rape—the children of mulattoes were also classified as mulattoes, for example, and not all sexual relations between white men and slave women were coerced—their unusually strong presence in the area, especially compared to regions such as lowcountry South Carolina (as will be discussed below), suggests that slave women with absent husbands were at risk of unwanted attention from white men.[17]

Many prominent scholars, including John Blassingame, have argued that enslaved men preferred cross-plantation marriages so that they would not have to witness their wives being beaten, overworked, raped, or otherwise abused during the week. This would have made slaveholding size irrelevant and may indeed explain some broad marriages, but most sources from Fairfax County (and the other counties chosen for this study) do not confirm this view. Indeed, the admittedly scant evidence from some of the larger plantations in the early decades of the nineteenth century provides enough examples of co-residential couples to suggest that slave men availed themselves of the opportunity to marry on the home plantation whenever possible, despite the abuses upon their wives that they may have had to witness. A group of runaways from Rose Hill plantation in 1804, for example, was described as a complete family consisting of Sam, his wife Suckey, their teenage daughter Jane, and "three small Children." At Sully plantation in 1809, one slave named John lived "with his wife Alice, and their eight children: Patty, Betty, Henry, Charles, John, Margaret, Milly and Frank." Austin Steward, who also grew up on a relatively large plantation around the turn of the nineteenth century, spent his early years in a "small cabin" which housed his "father and mother—whose names were Robert and Susan Steward—a sister, Mary, and myself." And Bushrod Washington, the master of Mount Vernon and more than eighty slaves, advertised in 1811 for a missing slave girl who "has a father & mother living at my farm."[18]

The 1810 estate inventory of William Fitzhugh, owner of Ravensworth plantation and with over two hundred bondspeople the largest slaveholder in the county, provides one of the only early-nineteenth-century sources from the area in which a majority of the slaves were listed in household groups, allowing for a limited analysis of domestic arrangements. Of the thirty-five households recorded in the inventory, at least eighteen consisted of co-residential couples (or the remnants of co-residential couples) with or without children, including ten of the thirteen households on the main Ravensworth farm of the estate (the plantation was divided into four "farms"). One household consisted of a single father: Noah lived with his six children, ages one to ten. Since slave children were legally the property of their mother's owner, it is certain that Noah's wife had originally lived with them but was since deceased (or sold). Seventeen households on the plantation were headed by what appear to have been single mothers, most of them residing on the smaller Backlick and Centre farms. Some of these women may have been widows (seven of them were over forty years old); a few appear to have had husbands who were listed as singles on other parts of the plantation (as will become clear below); still others may have had no permanent relationships or had husbands who lived on other holdings. George, the husband of one of these women, was advertised as a runaway in 1801, having last been spotted "at one of the Ravensworth Quarters, where he has a wife, and from whence he took away all his clothes."[19]

Finally, a total of sixty-eight slaves were listed and valued singly, a high number indeed. Other scholars who have analyzed the Fitzhugh inventory have dismissed these singles as kinless or not living in family groups, but this is highly unlikely. First, the Ravensworth estate had already undergone divisions and disruptions in the late eighteenth century, and so it is unclear how many families had already been ruptured. Second, slaves were often listed singly because they were intended to be sold or bequeathed separately from their family members, a common occurrence in Fairfax County. (Indeed, ten years later Fitzhugh's only son and main heir owned 158 slaves, a number that virtually matches the number of slaves listed and valued in household groups in the 1810 inventory. The "singles" listed may have therefore been sold or bequeathed to another family member.) Twelve "singles" were in fact children aged twelve or under—perhaps orphans, but certainly not kinless or living alone. Several singles were definitely related to each other or other families. John Bossee Senior (50), Sarah Bossee (50), and John Bossee Junior (28) were undoubtedly husband, wife, and son, yet they were all listed singly. Several more singles had the surname Bossee, and other slaves listed singly shared surnames as well, indicating that they were almost certainly family. Thirty-three singles were adult men (ages fourteen to forty-five), and many may have had wives living on other estates, but it cannot be entirely discounted that a few of these men even had wives on the same plantation. Aaron Clarke (30) and Joseph Clarke (30), both listed as singles attached to the main Ravensworth farm of the estate, may very well have been the husbands of Ailey Clarke (28) and Milly Clarke (30), who were listed as single mothers resident on Backlick farm—it is possible that these men were intended to be bequeathed apart from their families, or that they were temporarily living and working on the Ravensworth farm when local officials came to take inventory. The intricate web of family structure on Fitzhugh's estate may be impossible to fully unravel, but the number of (originally) co-residential households on the plantation was no doubt higher than it appears.[20]

The Fitzhugh inventory points to an increased tendency among slaves to establish co-residential families on larger estates, which suggests that such domestic arrangements were probably the ideal for most bondspeople. Indeed, the age difference between some spouses on Fitzhugh's plantation suggests that slave women sometimes may have preferred to marry substantially older men on the home plantation than men closer to their own age on neighboring slaveholdings. One slave woman named Nanny was married to a man eighteen years her senior. Another woman named Nelly was twenty-five years younger than her husband, Billy Douglass. It can be ruled out that Billy was in fact Nelly's father, because one of Nelly's sons was named Billy Jr.[21]

In the records of one relatively large slaveholder, B. R. Davis, naming practices may be the only clue available to ascertain slaves' family structures. Davis was the owner of seventy bondspeople when he died in 1823. The inventory of his estate lists sixteen able-bodied slave men, nine mothers with a total of thirty-five children, five women without children (one of whom was named after one of the mothers, and was thus undoubtedly her teenage daughter), and five elderly slaves. Although Davis's slaves are not listed in family groups, it is interesting to note that exactly six of the nine mothers had sons whose names matched the names of able-bodied men living on the plantation, suggesting that these men may have been the fathers. If so, an unusually high proportion of Davis's slaves lived in co-residential households. Moreover, two of the elderly slaves were married to each other.[22]

Toward the end of the antebellum period, large slaveholdings such as those owned by William Fitzhugh and B. R. Davis disappeared as slaveholding size shrank to farm proportions. Sources from the 1850s therefore provide fewer examples of co-residential couples; the few that do come mostly from slave testimonies. Francis Henderson, a refugee in Canada who told interviewers that he grew up on a plantation near Washington that had "about forty slaves on the place," lived in a co-residential family consisting of his father, mother, and twelve sisters. Carter Dowling, the slave of Maria Fitzhugh, who held more than twenty-five people in bondage, told abolitionists in the northern city to which he fled that "his mother, father, five brothers and six sisters [were] all owned by Miss Fitzhugh, [and] formed a strong tie to keep him from going." Carter thus grew up in a co-residential household, but the Dowling family was probably originally established on a holding that contained more than twenty-five slaves. Maria Fitzhugh inherited her slaves, and Carter's mother and father had most likely met and married while living on one of the larger Fitzhugh plantations that was later broken up—in this case keeping Carter's family together. Former slave Frank Bell, who claimed to have grown up on a relatively large plantation near Vienna, also lived and worked with his mother, father, and four brothers on the same holding. Bell's uncle Moses, however, who was the driver on the plantation, was forced to search elsewhere for a mate. And with slaveholdings in western Fairfax more scattered than in the vicinity of Alexandria, Moses eventually ended up marrying a woman who lived twelve miles away. Bell recounted that his uncle "was married to Aunt Martha, who was a slave on Parson Lipskin's farm, down near Hamilton Station, 'bout twelve miles from Vienna. Every Sunday Marser let Uncle Moses take a horse an' ride down to see his wife an' their two chillun, an' Sunday night he come riding back; sometimes early Monday morning just in time to start de slaves working in de field." In general, however, it appears that slaves in Fairfax County were more likely to form co-residential households and conjugal unions on large plantations, suggesting that that was probably the ideal, if an ever-diminishing possibility.[23]

Settling Down: Georgetown District, South Carolina

On the expansive rice plantations of the lowcountry, enslaved men and women encountered a different framework of boundaries and opportunities when it came to seeking a mate and establishing a nuclear family. In the previous chapter it was shown that slaveholdings in Georgetown District stood out in the antebellum South for the large numbers of bondspeople they contained. At their peak of wealth and power in 1850, the rice masters held 74 percent of the local slave population on holdings with more than one hundred slaves and 46 percent on holdings that contained more than two hundred slaves. Some estates controlled several hundreds of slaves; one counted over a thousand. Moreover, the sex ratio between men and women of reproductive age remained relatively balanced between 1800 and 1860. In short, life along the tidal estuaries of Georgetown District would appear to have been conducive to slave family formation.

In such a setting it seems reasonable to expect that enslaved people living on the antebellum rice plantations would have fared better than their counterparts in northern Virginia in establishing co-residential families and two-parent households. Indeed, even some of the demographic boundaries that slaves in Georgetown District encountered were probably not as formidable as they initially appear. Since many large rice planters owned multiple plantations, a large number of slaves no doubt found that the number of potential partners with whom they could establish a co-residential household extended beyond the number that lived on the home plantation. Joining a husband and wife from separate plantations but owned by the same master would have merely entailed a redistribution of labor, after all, not a purchase or sale. Moreover, one might expect that enslaved people's prospects for co-residential family formation would have actually improved over time as a result of the steady growth of slaveholding size. As the following shows, an unusually high proportion of enslaved people in Georgetown District did in fact succeed in establishing co-residential families throughout the antebellum period, providing a very different picture from that in northern Virginia and further supporting the theory that such arrangements were the ideal for most slaves throughout the antebellum South.

Most of the probate records for Georgetown District were destroyed during the Civil War, and so estate accounts from local slaveholders are rare, especially for the early decades of the nineteenth century. The lack of official records are in some measure compensated, however, by the number of private documents left by masters and overseers who lived and worked on the rice plantations. Since plantations in the lowcountry were such large-scale enterprises, their operations were generally better documented than the small farms of northern Virginia. Slave inventories, account books, ration books, and records of births and deaths were all carefully kept by local slaveholders, and many of these have survived. Moreover, slaveholders and overseers in the lowcountry often recorded family connections among their slaves, if only to facilitate keeping track of such large groups of bondspeople. Consequently, the marriage strategies and family structure of slaves living on many plantations can be ascertained with unusual accuracy. Virtually all sources indicate a strong presence of co-residential slave families.

For the early antebellum period, the plantation book left by the McDowell family—which roughly covers the first four decades of the nineteenth century—provides an illuminating case in point. Davison McDowell, who inherited his father's slaves and ran the plantation until his death in 1837, was a relatively large rice planter, like most of his peers; in the census schedules of 1830 he was listed as the owner of 111 slaves. Although McDowell's slaves are nowhere listed in family groups, he did meticulously record his slave births into his plantation book, documenting not only their names and years of birth but also their parentage. Of the seventy-six slaves born on the McDowell plantation between 1800 and 1830 (not all of whom survived), no less than sixty-three (83 percent) were born to co-residential parents, an unusually high proportion. Many co-residential couples had several children over a long period of time, suggesting remarkably stable relationships. Linda and Jack, for example, had seven children; Fena and Southerland, and Juno and Manza, each had six. Only eleven slaves (14 percent) were born to single mothers; the parentage of two slaves born on the plantation was not recorded at all. In short, from the perspective of slave children born on the McDowell estate, the chances of growing up in a two-parent household were very high indeed.[24]

The inventory of slaves belonging to the estate of Benjamin Alston Jr., in 1819 indicates a similar prevalence of co-residential couples and two-parent households. Of the 132 slaves listed in Alston's inventory, 113 were grouped into one of seventeen family units, each with a collective value. Fourteen of these families (containing a total of 108 slaves, or 82 percent) were headed by co-residential couples. One of these families consisted of an adult man named Jacob and his son Will, suggesting that Jacob's co-residential wife was probably deceased. Only three families were undisputedly headed by single mothers, but a few of the multigenerational families headed by co-residential couples may have in fact been hiding some single mothers among their offspring. Prince ($1) and Venice ($150), most likely elderly, are listed with what appear to be their two daughters and their daughters' children from possible broad marriages: Grace ($800) and two children named Affee ($200) and Pierce ($100); and Diannah ($800) and her children, Prince ($200) and Cupid ($100). A total of nineteen slaves in the inventory were listed singly, eighteen of whom were men or boys, and two of whom were valued at only $1, suggesting that they were either elderly or otherwise physically unfit for labor. It is unclear whether any of these single men were married to women on other plantations.[25]

The records of other plantation owners from the early nineteenth century confirm the trend. A majority of the slaves residing on the plantation of William Lowndes lived in two-parent families in 1813, accounting for 163 of his 256 slaves. The slave inventory taken of the Paul D. Weston estate in 1837 points to an unusually high percentage of two-parent households among the slave population. Of the thirty-one households listed on his plantation, no less than twenty-four (containing a total of 127 slaves, or 80 percent) were headed by co-residential couples. Only seven households were headed by single mothers (27 slaves, or 17 percent), and four male slaves were listed singly.[26]

By the late antebellum period, enormous plantations that held vast numbers of bondspeople had become the norm in Georgetown District, and plantation records from the last decades before the Civil War broke out provide plenty of evidence that co-residential couples and two-parent households were numerous in local slave villages. On Weehaw plantation, which contained 265 slaves in 1855, sixty-seven of sixty-nine households were headed by co-residential couples. At White House plantation, owned by Julius Izard Pringle and containing 110 slaves in May 1858, fifteen of seventeen households were headed by co-residential couples (although three young single mothers were listed under their co-residential parents' names).[27] On Fairfield plantation 81 percent of the slaves lived in two-parent households in 1841; on Dirleton in 1859 it was 58 percent; on Northampton in 1853 such households accounted for 86 percent of the slave population; on True Blue 73 percent; and on Paul Weston's estate in 1860 some 80 percent of the slaves lived in households headed by co-residential couples.[28]

Not only plantation records but other qualitative sources indicate that co-residential marriages were widespread on the vast estates of late antebellum Georgetown District as well. The Reverend Alexander Glennie, who was employed by several rice planters in the Winyah Bay area (especially on the Waccamaw) to preach to the slaves, recorded in a parochial report in 1860 that the practice of joining enslaved men and women "together in matrimony" was simply "easier to carry out . . . on large plantations in the country, where the parties in almost all cases belong to the same owner." Even local runaway slave advertisements provide evidence of co-residential family structure, an interesting contrast to the situation in Fairfax County. One such advertisement, which ran in the *Winyah Intelligencer* in March 1819, announced the flight of an African slave named Fortune, his American-born wife Dinah, and their infant child. Another planter living on the Black River advertised the escape of his "negro man named July, who at the same time carried with him his wife Dolly, and their children, Tom, Lizzy, Judy, and Eleanor." In 1830 a carpenter named George, owned by one Mrs. Murray, absconded with his daughter Stellah, a field hand. Again, since children were legally the property of their mother's owners, it is almost certain that George's wife (and Stellah's mother) had originally lived with them but was since deceased.[29]

Finally, former slaves from the district recalled growing up with both parents and even co-residential grandparents on the same estate. Sabe

Rutledge related to interviewers of the Federal Writers' Project that his father and mother, and his grandfather and grandmother, all lived on the plantation where he lived as a child. Gabe Lance, born and raised on Sandy Island, told interviewers that his "Pa and Ma" were owned by the same master, and that his grandfather was the driver on his plantation—again, a resident grandfather living with any of his grandchildren meant that he had married a woman from the same plantation. Ben Horry grew up in a two-parent household: "Father dead just before my mother. They stayed right to Brookgreen plantation [owned by J. J. Ward] and dead there after they free." William Oliver told interviewers that his "Father Caesar Oliver; Mother Janie . . . Had four brother, twelve sister" on his plantation. And Margaret One lived with her mother and father, the latter of whom was the "principal plantation carpenter" on Wachesaw Plantation on the Waccamaw River.[30]

Not all slaves in Georgetown District succeeded in establishing two-parent households, however. Those who lived on holdings that contained fewer than one hundred slaves (an average of 42 percent of the slave population between 1800 and 1830, although many of these lived in village centers and not on plantations) would have been harder put to find spouses on the home estate. And while evidence suggests that slaves living on the larger plantations found more opportunities to marry resident partners, every plantation counted a number of bondsmen and women who were forced by the demographic boundaries of the estate to search elsewhere for mates.

As stated above, eleven of the children born between 1800 and 1830 on the McDowell plantation were born to one of nine single mothers. The absent fathers of some of these children were specifically recorded: James, born in 1828, was the "son of Lucy and Robert F. W. Allston's James"; Priscilla, born the same year, was the "daughter of Maria and Henry D. Shaw's Joe"; Columbus was the "son of Libby and W. Coachman's Branch"; and Marcus was the "son of Molly and Dines Oxen (free man)." Fourteen of the women living on Charles C. Pinckney's plantation in 1812 also had children by men from different holdings. One of these women, Moll, had had relationships with three different men in succession, all living on different holdings. The father of Moll's teenaged son named Sam was "Gen. Rowler's Jared"; but she also had three adolescents by "Wm. Elliott's Downy," and a small boy (between two and five years old) named February, whose father was "Mr. S. A—'s Davy." Benjamin Alston counted three female-headed households and nineteen singles among his slaves in 1819; William Lowndes counted fourteen mothers with absent husbands and forty-nine singles; and Paul D. Weston owned seven slave women whose husbands lived elsewhere, as well as four single males.[31]

Even in the late antebellum period, when slaveholding size had reached its peak, a minority of slaves continued to establish cross-plantation families. At Weehaw plantation in 1855 two out of sixty-nine family units consisted of female-headed households, and twenty-eight slaves (out of 265) were listed singly. Approximately 19 percent of the slaves living on Friendfield plantation in 1841 were either singles or residing in female-headed households. At Northampton, such slaves accounted for 14 percent of the population. Even Ben Horry, owned by grandee J. J. Ward on the Waccamaw Neck, claimed to have courted a couple of Benjamin Allston's slave girls as a young man. "FUSS one I go with name wuz Teena. How many girl? Great God! I tell you! FUSS one Teena; next Candis. Candis best looking but Teena duh largest! Go there every Sunday after [Sunday] school. Oatland plantation—belong to Marse Benjamin Allston. Stay till sunset. Got to have paper. Got to carry you paper." Neither girl became Horry's wife, however, and it is unlikely that many of Ward's one thousand plus slaves resorted to broad marriages.[32]

On some plantations, however, broad marriages were more numerous, a reminder that the demographic boundaries with which slaves of different owners were confronted were far from uniform. At Dirleton plantation in 1859, some 42 percent of the slave population did *not* live in two-parent households, although the abroad husband of one woman named Rose was eventually purchased and united with his wife, perhaps through their negotiation. Of the thirty-two children born on the plantation of Charles C. Pinckney between 1840 and 1854, fourteen were born into female-headed households (almost 44 percent). Most of the broad relationships on Pinckney's estate were specifically recorded: Rose was born to "Binah and Dr. Stuart's Marcus" in 1841, for example, and Sary (born in 1852) was the daughter of "Sophy and Mr. EBL's Richard." But although the number of slaves who lived in divided households of course varied from plantation to plantation, it was almost everywhere consistently lower than the number of slaves who lived in co-residential families.[33]

Indeed, a preference for co-residential domestic arrangements among the slave population in Georgetown District may have largely protected women from unwanted sexual advances by white men, at least compared to the situation in Fairfax County. Slaves on the rice plantations lived largely isolated from the white population anyway, but even the white overseers appear not to have often dared meddle with slave women who lived with their husbands. In 1850 the mulattoes of Georgetown District amounted to only 0.5 percent of the slave population; in 1860 they formed 1.3 percent of the population. Many were probably the offspring of mulattoes and not direct products of rape, but compared to 22 percent and 30 percent of mulattoes in Fairfax County in 1850 and 1860, respectively, slave women were obviously not at great risk in Georgetown. Most of the mulattoes there were domestic servants who lived in the county seat and came into the most contact with whites (as many as sixty-five of ninety-nine mulattoes in 1850, for example, lived in the city of Georgetown). On plantations, mulattoes were rare and often the products of interracial contact between white men and the household slaves who waited on them. Ben Horry recalled that one domestic servant named Susan "had three white chillun. Not WANT em. HAB em. Boy near bout clean as them boy of Missus! Tief [Thief] chillun show up so! Woman overpower!" Co-residential family structure among most slaves, however, may have prevented such incidents from occurring as frequently as they did in Fairfax County.[34]

For those who could not find—or chose not to marry—partners on the home plantation, however, mobility played a crucial role in courting practices and family formation. Slaves were not permitted to be absent from their plantations without written passes, but antebellum sources indicate that these were generally granted upon request (pending good behavior and as long as their absence did not interfere with work). David Doar, an antebellum rice planter on the Santee, recalled that "those men or women who had wives or husbands on adjoining places were, of course, given a pass to visit them." Plowden Weston likewise ordered his overseer to uphold the rule that "no one . . . be absent from the place without a ticket, which is always to be given to such as ask for it, and have behaved well." Neither planter specified that these passes were to be given exclusively on the weekends; perhaps this was such general practice that no further specification was necessary. Evidence from elsewhere in the lowcountry, however, suggests that those slaves who managed to finish their tasks by the early afternoon might sometimes have been afforded the opportunity to visit their girlfriends or wives on other holdings during the week. Sam Polite, born on St. Helena Island of a cross-plantation marriage, told WPA interviewers that "w'en wuk done, [a slave man could] visit his wife on odder plantation. Hab pass, so patrol won't git um."[35]

Some late antebellum sources indicate that many lowcountry planters only reluctantly granted their slaves permission to court across plantation lines, preferring for their bondspeople to find spouses on the home plantation whenever possible. Sam Mitchell, who was enslaved on Lady's Island, recalled that his master "didn't like for his slaves to marry slaves on another person's plantation, but if you did that then you had to get a pass to visit your wife." Frederick Law Olmsted reported of a rice planter in the Georgia lowcountry that "Mr. X does not absolutely refuse to allow his negroes to 'marry off the place' . . . but he discourages intercourse, as much as possible, between his negroes and those of other plantations; and they are generally satisfied to choose from among themselves." Such evidence suggests that as opportunities to establish co-residential families in the lowcountry increased, they not only remained the ideal arrangements among most slaves but among slaveholders as well, who no doubt felt uneasy about allowing their slaves to move about the landscape unsupervised during their free time. Even slaveholders could not ignore the demographic boundaries of their plantations, however, which always left some enslaved men and women without potential mates.[36]

It is interesting that many slaves in Georgetown District who courted or married across plantation lines stayed within the immediate vicinity of their home estates, usually finding partners who lived on neighboring plantations. Considering the distances between slaveholdings in many areas of the district, choosing a spouse who lived as close as possible would have been especially advantageous, as it would have reduced travel time and increased families' time to spend together. Doar specifically referred to slaves who had "wives or husbands on *adjoining* plantations," as if that was the most obvious choice for slaves who sought to marry across property lines. The slave mothers whose cross-plantation marriages were recorded by Pee Dee planter McDowell also testify to this tendency. Lucy bore a child by Robert F. W. Allston's slave James, who at the time lived "next door" to the McDowell plantation, just over a mile away by river. Likewise, Libby bore a child by W. Coachman's slave named Branch, who lived two plantations to the north (approximately three miles by river). Plowden Weston's rules required "all persons coming from the *Proprietor's other places* [to] show their tickets to the Overseer, who should sign his name on the back." The latter suggests that the slaves of planters who owned multiple plantations often courted slaves who lived on the master's other (usually adjoining) estates, no doubt because the chances of establishing a co-residential household were higher in such cases.[37]

Interplantation visiting created a web of social networks that connected virtually all slaves to some extent, as even members of co-residential families requested passes to visit with friends and extended kin. For some, however, such networks provided the lifeline between themselves and their spouses and children who lived beyond the boundaries of the home plantation. Limited evidence of broad social networks can be found in some of the runaway slave advertisements for the early nineteenth century. In 1819 a carpenter apprentice in Georgetown village absconded and was presumed to be hiding out "on Mr. Kinloch's plantation, where he has a wife." In May 1829 the owner of a vessel docked at Georgetown advertised the flight of his slave named Paulladore, who "has a wife on Mr. H. D. Pinckney's plantation near Georgetown, where it is supposed he will be harbored." John Allston advertised for the apprehension of his slave Adam, whom he had bought at an estate sale on the Santee, adding that "he has relations at Dr. Lynch's Peach Tree place (South Santee), and at Miss Bowman's, North Santee, and is generally acquainted in Georgetown District." It is unclear whether Adam had a wife or girlfriend on the plantations where he was suspected to be hiding, but such cases illustrate the extent of social networking between neighboring slave communities. Enslaved men and women with few or unattractive prospects for marriage on the home plantation tapped into such networks in search of a mate.[38]

Such slaves appear to have been in the minority, however. Throughout the nineteenth century in Georgetown District, numerous enslaved men and women were afforded opportunities to marry partners on the home plantation and establish co-residential and two-parent households—opportunities they appear to have actively seized.

Unequal Opportunities: St. James Parish, Louisiana

Marriage strategies and family structure among enslaved people living in St. James Parish reflect the skewed demographic consequences of the nineteenth-century sugar boom. The development of slaveholding size in the parish differed from that in Fairfax County and Georgetown District. Cane cultivation generated fantastic wealth and stimulated rapid growth in slaveholding size, but in many ways southern Louisiana still found itself in the developing stages of a plantation economy. As stated in chapter 5, astronomically high starting costs in the sugar industry forced many would-be planters to begin operations with a limited number of slaves, gradually expanding their labor force after they had produced a few respectable crops and paid off some of their heavier debts. Some failed and went under. In 1810, with the sugar industry still in its infancy, a majority of slaves in St. James lived on farms that contained twenty slaves or less; in 1860 a quarter of them still did. On the other hand, local planters who succeeded in expanding their operations during the course of a few decades, or who had access to substantial capital to begin with, such as Valcour Aime, Benjamin Winchester, John Burnside, and most of the joint ventures, were able to start operations on a larger scale and quickly achieve the status of grandees by the outbreak of the Civil War. By 1860 some 40 percent of the enslaved population lived on holdings that contained more than one hundred slaves (9 percent lived on holdings with more than two hundred slaves).

It seems reasonable to expect that especially in the early decades of the nineteenth century, finding a spouse on the home estate may have been difficult for many slaves, as slaveholding size was moderate and the sugar industry still in its infancy. As the sugar boom gained momentum in the late antebellum period, however, one might expect opportunities for co-residential domestic arrangements to have increased with time for most slaves, although disproportionately more for women than for men. The latter faced the greatest challenge in establishing formal families at all, whether co-residential or not.

In chapter 5 it was shown that in 1820 some 59 percent of enslaved people between the ages of fourteen and forty-five were men, leaving at least 344 men of reproductive age without potential mates in that year; by 1850 the parish counted 61 percent male slaves, with the aggregate male surplus amounting to 914. Sugar planters preferred strong, able-bodied men to perform the exhausting work of cane cultivation, and their purchasing habits quickly resulted in a male majority. Certainly some early planters with wiser business instincts than their peers recognized the long-term advantages of purchasing more women, which would encourage family formation and increase the slave labor force naturally. The Duparc brothers and their business partner Locoul, for example, felt that something needed to be done in the 1820s about the demographic imbalance on their jointly owned and operated plantation, which contained sixty men and thirty-two women. They purchased thirty slave women to ensure an adequate future labor force, and, according to Laura Locoul Gore, who grew up on the plantation, "by the 1840s, *Duparc Frères & Locoul* was awash in slave children." Their neighbors were less insightful, however, and the male majority remained a structural problem.[39]

Throughout the antebellum period, family formation was thus consistently truncated for about one-third of the male slave population. Bound to plantations that lacked a sufficient number of potential mates and surrounded by slaveholdings with similar demographic compositions, enslaved men in the sugar parishes competed fiercely for the attention of the limited number of available women, a situation that not only put young women under considerable pressure but also inevitably led to conflicts between male rivals. One historian argued that drivers on the sugar plantations were often forced to intervene "to prevent possible trouble between two rivals for a likely looking girl." Francis and Theresa Pulszky, who visited the sugar parishes in the 1850s, were told by one slaveholder that fighting was a major problem among slave men: "They would stab one another, if we did not interfere with the whip."[40]

While many men were probably informally "adopted" into the households of others, establishing extended networks based upon fictive kinship, most lamented their failure to marry. When correspondent Frederick Law Olmsted, who visited the sugar parishes in the 1850s, asked a thirty-three-year-old slave who had never been married what he would do if he was freed from bondage, the man replied that "de fus thing I'd do, I'd get me a wife; den, I'd take her to my house, and I would live with her dar." When freedom finally came, thousands of men—confronted with few prospects for marriage in their home parishes—took to the road. Historian John C. Rodrigue, who extensively studied the transition from slavery to freedom in the Louisiana sugar parishes, found that "the apparent lack of familial attachment among many male slaves had important repercussions after emancipation: the preponderance of single, unattached young men would contribute to what planters viewed as the unstable and transient character of their labor forces under free labor."[41]

Indeed, one could argue that the large numbers of unattached men in the sugar parishes contributed to the relatively unstable and transient character of their labor forces under slavery as well, as revealed by a perusal of the frequent runaway slave ads from the region. Unlike in Fairfax County, where runaways and truant men were usually suspected of lurking about near their wives' places of residence, cross-plantation marriages are almost never recorded in the advertisements for slave men from St. James Parish, probably because these men were not married. In June 1855, five men (ages nineteen to thirty) ran away from Samuel Fagot's plantation, and another twenty-five-year-old man named Ned absconded from G. F. Roman's estate. None were suspected of hiding out near another plantation where they might have had wives. Former slave James Conner, who "saw some hard times on the [sugar] plantation" from which he escaped in 1857 at the age of forty-three, did not mention to northern interviewers that he had ever been married or broken any family ties when he fled Louisiana. Such cases are fairly typical for the region.[42]

As in northern Virginia and lowcountry South Carolina, however, evidence from large and moderate-sized plantations in St. James Parish indicates that enslaved men who were fortunate enough to find a mate among the local slave population often availed themselves of the opportunity to marry women from their own plantations, establishing co-residential families whenever possible. For the opening decades of the nineteenth century, when the sugar industry was still developing and slaveholding size was more moderate, evidence of slave family structure is unfortunately scarce; most slave inventories from this period simply list slaves' names and respective values, indicating no family ties. Still, in the fragmentary sources that are available in the early probate records, evidence can be found of co-residential couples living on plantations of various sizes.

The records of Michel Cantrelle, who with sixty slaves was the fourth-largest slaveholder in the parish in 1810, mention a number of co-residential slave families. In February of 1810 Cantrelle's records mention a "famille d'Esclaves" which he had acquired for the bargain price of $655 from F. Blanchard. Other miscellaneous papers record the collective values of two co-residential slave families living on the plantation in 1810: Cesar, Helene, and their four children ($1500); and Philipe, Jacinta, and their three children ($1000). Elsewhere in the Cantrelle estate records are notes that in 1816 one Bartholome Congo (aged 30) was married to Maria Congo (35). It is uncertain whether this couple had children.[43]

Cantrelle was a relatively large slaveholder at the time, but surprisingly a number of two-parent households appear in the records of even modest estates in St. James Parish. Due to the large number of imports and purchases on even small and moderate-sized slaveholdings, opportunities to establish co-residential families were probably better in the sugar country than on farms and plantations of a similar size in Virginia, because fewer slaves would have been related to each other. The slave population in St. James shot up by 76 percent between 1810 and 1820, and again by 63 percent in the decade that followed. Predictably, by the 1830s there is more evidence of two-parent households among the slave population. On the relatively small plantation of sugar planter Daniel Blouin, who owned thirty-four bondspeople upon his death in 1831, a number of examples can be found of co-residential couples living with their minor children. Augustin (aged 35) and Adzire (33) lived with their four youngest children (ranging in age from two to nine). Likewise, Bazile (35) and Petty (40) resided with their six youngest children (ages one to nine). The succession papers of Joseph Melancon, who owned thirty slaves in 1837, contain evidence of at least three co-residential slave couples living with their children. Interestingly, all three of these couples were intercultural: Levoille was a Louisiana-born slave ("creole") married to Marie, who had been imported from the Congo; William, an "American" (imported from one of the Anglo-American slave states), was married to Finne, a creole; and Abraham, also an "American," was likewise married to a creole woman named Jemima.[44]

From the third decade of the nineteenth century on, sources that indicate co-residential family structure among the slave population are far more numerous. In 1840 some fifty-one slaves were advertised at the estate sale of one deceased planter; almost half of these slaves were single men (twenty-four), but nearly all of the rest (at least twenty-two) were living in co-residential family units, including one widower named Ben (aged 50) living with his son Petit Ben (20). George Mather, who owned 113 slaves upon his death in 1837, was one of the largest planters of the parish. The probate records concerning the settlement of his estate list most of these slaves in family groups, a majority of which were headed by two parents. Thirty adult men and women were listed as co-residential couples, and a total of forty slaves (children and young adults) were described as the offspring of these couples. In other words, seventy slaves out of 113 (62 percent) were living in co-residential family groups, although this does not include the single adult sons of some couples, nor some of these couples' grandchildren by a few unmarried teenage daughters.[45]

By the final years of the antebellum period, when local slaveholdings had reached their greatest size, co-residential families in St. James Parish were clearly widespread. In 1852 the sugar plantation of grandee P. M. Lapice contained 256 slaves, of whom at least 156 (or 61 percent) were living in two-parent households. A total of twenty-five families are listed in the 1860 slave inventory of W. P. Welham's plantation, which contained 125 slaves, and no less than twenty of these families were headed by co-residential couples. Of the thirty-four children under ten years of age who were listed in an 1855 slave inventory of Constancia plantation, owned by Samuel Fagot, at least twenty-five (74 percent) were born into two-parent households. Buena Vista plantation, owned by Benjamin Winchester and later run by his widow, also contained a majority of two-parent households among its slaves. Records from the Freedmen's Bureau in 1865 contain evidence of at least twenty-one households that were headed by two parents or widowers, accounting for ninety-nine slaves (62 percent), all of whom had been resident on the estate prior to emancipation. Wilton plantation, which contained 152 slaves upon emancipation in 1865, contained forty-one households headed by two parents or widowers, three female-headed households, eighteen single men, and three (childless) single women. Even on the moderate-sized Webre plantation, where only thirty-one adult slaves were resident upon emancipation, eighteen consisted of co-residential couples; fourteen were single men and two appear to have been single women.[46]

Historian Ann Patton Malone estimated that between 1810 and 1864 an average of 49 percent of slaves in the Louisiana sugar parishes lived in two-parent households. The individual plantation records of some slaveholders in St. James Parish suggest perhaps an even higher average, but other qualitative sources confirm that co-residential arrangements were at least widespread on the sugar plantations in the late antebellum period. The overseer on the enormous Houmas plantation, which straddled the border between St. James and Ascension parishes, regularly recorded evidence of two-parent households, with entries in the plantation diary such as "Sam's family quite well" [after illness]; and "Grace Nott had twins last night. . . . I call them Peter and Grace after their mother and father." Travelers also frequently mentioned the existence of co-residential families. When Francis and Theresa Pulszky asked an enslaved woman on the plantation they visited whether she was married, the woman replied: "Oui, mon mari est dans les champs." William Howard Russell, upon entering the cabin of one enslaved man on Alfred Roman's plantation in St. James Parish, found the man's wife "smoking a pipe by the ashes on the hearth, blear-eyed, low-browed, and morose," as well as the co-residential couple's young daughter, "some four years of age." While visiting one of the Burnside plantations, Russell spoke to an elderly African slave named Boatswain, "who lived with his old wife in a wooden hut close by the margin of the Mississippi."[47]

Finally, former slaves from the sugar country overwhelmingly claimed to have grown up in two-parent households. Fred Brown grew up on a plantation on the west bank of the Mississippi which contained his "pappy, mammy and three brudders and one sister, Julia, and six cousins." Catherine Cornelius recalled to interviewers that on the cane plantation where she was raised "my ma and pa and de whole family was dere too." Gracie Stafford, born enslaved on Myrtle Grove plantation in St. James Parish, related that her entire family had been owned by White and Trufant; after the war, her "ma and pa took us children to Grand Prairie, and that's where I growed up." Elizabeth Ross Hite claimed that her co-residential "mother and father came from Richmond, Virginia," raising two sons and two daughters in Louisiana. She added that "some of dem darkies didn't care about master, preacher, or nobody. They just went and got married, married deyselves." Daffney Johnson said that while her mother and father lived together and had children on the plantation to which they were bound, they never had any kind of wedding ceremony: "They never did marry like de people do now'days."[48]

Two-parent households were clearly widespread in St. James Parish by the end of the antebellum period, yet not all slave families were established by co-residential couples. During the early decades one might presume that cross-plantation arrangements would have been common, as slaveholding size in the sugar parishes never approached the grandeur of holdings in the lowcountry; the proportion of slaves living on sugar "farms" containing twenty or fewer slaves ranged from a high of 62 percent in 1810 to a low of 26 percent in 1860. However, the surplus of adult males on most holdings, even small holdings, along with the high percentage of imports, seems to have limited the necessity for cross-plantation marriage among most enslaved women. These last two factors, which did not play a significant role in either Fairfax County or Georgetown District, may explain why clear indications of broad marriages in St. James Parish do not appear in the records with as much frequency, although they were far from absent. Ann Patton Malone estimated that an average of 24 percent of slaves in the sugar parishes lived in cross-plantation households; for St. James this estimate may be valid, but evidence is admittedly fragmentary.[49]

In various slave inventories of local slaveholders throughout the nineteenth century, limited indications of what appear to be single mothers can be found, but for the early decades of the antebellum period, sources are often vague and misleading. Succession papers frequently list women and children together with no explicit connection to any resident men, but as marriages between slaves were rarely recorded in these inventories it is possible that these women were indeed married and living with their husbands. Their husbands may have also been deceased, which—considering mortality rates—would not have been unusual even for young parents.[50]

Inventories dating from the latter decades of the antebellum period contain better evidence of cross-plantation relationships (although this does not indicate that such arrangements were more prevalent then than in the early nineteenth century). In the 1837 estate inventory of George Mather's plantation, whose 113 slaves were listed in family groups, eighteen small children (under the age of ten) were listed under one of eleven single mothers. Five of these mothers, however, were in fact young women under the age of twenty who were still living in their parents' households—their babies were likely the products of premarital sexual activity (probably with resident young men) rather than broad marriages. For instance, the household of Raphael and Rachel included five teenage daughters, three of whom had babies. One of them, Patsey, was only thirteen years old. Historian Richard Follett has argued that teenage pregnancies in the sugar country were especially prevalent, possibly as a result of increased sexual pressure from the predominantly male community. Casual sexual liaisons between unmarried men and women may have helped to alleviate tensions among the male majority; some women indeed chose not to marry at all, preferring a more permissive sexual life instead. On Mather's plantation the other single mothers were adult women who appear to have been maintaining stable relationships with men from other holdings, however. Suckey, a thirty-year-old woman who also lived in the same cabin as her mother and father, had three children ranging in age from eight to six months old. Thirty-five-year-old Benedicté had five children, aged eighteen months, six, nine, fifteen, and twenty. And Betsy, aged forty-five, had five children between twelve and twenty-five years of age. With such steady childbearing it is likely that all of these women were married to men from other holdings.[51]

Plantation records from the final years before freedom also indicate some broad marriages. The plantation of P. M. Lapice counted approximately eleven female-headed households, despite a large number of unattached resident men. Such households accounted for forty-three slaves, although some of these women were probably widows and not necessarily married to men from other holdings. Two of them were in their seventies and living with their adult sons, another was fifty, and a fourth was forty-four. Samuel Fagot's Constancia plantation counted five female-headed households in 1855, accounting for a total of nine children (or 26 percent of the children born on the estate). On the Welham plantation in 1860, five of twenty families (accounting for 17 percent of the slave population) were headed by women whose husbands did not live on the estate. (Zoé, a thirty-six-year-old mother of six, had a free black husband, a carpenter named Jean. For her, the advantages of having a free husband no doubt outweighed the advantages of having a co-residential slave husband.) Two of these women, Clemence and Eliza, had husbands on the Callouet plantation; Victoire's husband was not mentioned (she had recently died, leaving her five children orphans); and Phine (a childless woman) had a husband named William on the LeBourgeois estate.[52]

A handful of other sources also provide evidence of female-headed households and broad families in the sugar parishes. A letter to planter Jean Baptiste Landry from a man named George Jones, who was temporarily hiring one of Landry's slaves, informed the slaveholder that "your boy Scipio, who has a wife at the McCollam place . . . has been accused of improper conduct recently, which renders it improper for him to [go] to [that] place until it is explained." Victor Tixier recorded during his visit to St. James Parish in the early 1840s that "it often happens that a Negress has for a husband a slave of a neighboring plantation." Former bondswoman Annie Flowers told interviewers that she was "just a small child when my mother was set free." She did not mention having a father on the estate, although perhaps her mother was a widow or had never been married. Steven Duncan also failed to mention a resident father, recalling simply that when the Yankees came to liberate the slaves his "mother was in the quarters." And Solomon Northup, who was regularly hired out to the sugar plantations of southern Louisiana, claimed that in the sugar country there were many cases in which "the wife does not belong to the same plantation with the husband."[53]

For those enslaved people in St. James Parish who courted and married across property lines, transcending the geography of containment was regulated by the pass system, as it was in the rest of the South. At the Minor plantation in Ascension Parish, a typical estate for the region, prospective overseers were instructed "not [to] allow the negroes to go off the plantation without a pass nor must he allow negroes to come on the place without a pass, & must require them to present themselves & passes to him before going into the quarters or anywhere about the place." Another planter wrote of one of his slave men who had a broad family that he had "no objection to the Boy visiting his wife once a week." Former slave Hunton Love, who was born in 1840 and was a grown man by the time the Civil War broke out, recalled to interviewers that he and his fellow bondsmen "didn't leave the plantation often. . . . But if we did go, we had to have a pass or we'd be taken up. They was strick in those days." And Solomon Northup wrote that slave men who had wives on other estates were usually "permitted to visit her on Saturday nights, if the distance is not too far."[54]

As in Georgetown District, slaves in St. James who did marry across property lines sought partners who lived close by, usually on adjoining plantations. Tixier reported that slave women in St. James often had husbands who lived on "*neighboring* plantations." Unlike their counterparts in the lowcountry, however, slaves in St. James Parish enjoyed the advantage of living on such narrow estates that travel distances between neighboring slave quarters were probably reduced to a matter of minutes. The broad marriages recorded in the above-mentioned Welham plantation inventory provide a case in point. Clemence and Eliza both had husbands who lived on the estate of J. B. Callouer, who according to Persac's meticulously surveyed map of the Mississippi River plantations in 1858 lived only a few acres to the north, literally next door, on a narrow sliver of land. Phine, the childless slave woman, found her husband William living on Belmont plantation, owned by the LeBourgeois family and located immediately to the south—again, at a stone's throw. Indeed, so close were the slave villages of neighboring plantations that night visiting, though forbidden, occurred frequently. J. Carlyle Sitterson, in his study of the sugar plantations in southern Louisiana, found that during the week, when permission for men to visit their wives on other plantations was denied them, "they went anyway." On the McCollam plantation in neighboring Lafourche, even the women went night visiting. In 1845 the mistress of the estate put a chain around the ankle of one woman named Esther, to prevent her from "running about at night."[55]

While interplantation visiting may have been common, little evidence suggests that cross-plantation marriages overwhelmingly prevailed in St. James Parish or elsewhere in the sugar country during the antebellum period, which supports the research of Ann Malone. The proportion of mulattoes in the slave population—a possible indicator of absent fathers—cannot be accurately calculated from the census schedules because most slaves in the parish had been imported from regions such as Virginia, where mulattoes were already substantial in number. Single women were always at risk of course, and a few sources do reveal instances of miscegenation, perhaps partly due to the absence of husbands. During a visit to John Burnside's plantation William Howard Russell noticed that some of the slave children were "exceedingly fair," while the overseer "murmured something about the overseers before Mr. Burnside's time being rather a bad lot." Yet it does not appear that such cases were as prevalent in St. James Parish as they were in Fairfax County, perhaps because co-residential households in the former region were more numerous.[56]

Conclusion

Throughout the South in the antebellum period, enslaved people married and established simple families, often going to great lengths—under difficult circumstances—to do so. Yet their marriage strategies and family structures differed across time and space, depending on the types of demographic and geographic boundaries and opportunities with which they were confronted. Cross-plantation marriages and broad families could be found in each of the three regions chosen for this study, for example, but they were especially prevalent in northern Virginia, where opportunities for slaves to marry partners on the home place were severely limited, and indeed gradually worsened over time. Likewise, co-residential couples and two-parent households could be found throughout the non-cotton South, yet they appear to have prevailed in regions such as the South Carolina lowcountry and the Louisiana sugar country, where opportunities to marry resident mates were more plentiful. Finally, while each region counted singles among its respective slave population, only in St. James Parish did the demographic composition of the slave labor force consistently prevent substantial numbers of enslaved men from establishing nuclear families at all.

The marriage strategies and family structures of slaves living in Fairfax County, Georgetown District, and St. James Parish thus differed by degrees. Yet the nature of family formation in each region suggests that co-residential households were the ideal domestic arrangements for most slaves. They were not feasible for all bondspeople, however, and where they were not, enslaved men and women strove to adapt themselves accordingly, establishing nuclear families that transcended the geography of containment. How stable these families were in the long run, and how susceptible they were to forced separation, is the subject of the next chapter.

7

Forced Separation

Family formation among slaves did not guarantee that families would always remain intact, whatever their structure, whether co-residential or not. No slave family in the antebellum South was completely safeguarded from the prospect of forced separation, but scholars have long disagreed about the extent of family breakups and even the methods of forced separation. Much of the debate has centered around long-distance sale. Fogel and Engerman, for example, argued that long-distance sales accounted for only 2 percent of forcibly dissolved slave marriages in the antebellum South, whereas Michael Tadman has put the estimate at one in three. The domestic slave trade did not form the only threat to slave families, however—estate divisions, local sales, and long-term hiring also disrupted domestic arrangements and the nature of family contact. These factors have recently been explored by scholars such as Emily West and Wilma Dunaway, but regional differences in the threat of family breakups have yet to be examined in detail.[1]

Time and place mattered with respect to all of these methods of forced separation, and upon closer inspection it appears that families in particular regions were more at risk than others. The specific labor demands and nature of slaveholding in various localities could make forced separation a vague yet real possibility, or a constantly recurring nightmare. In an attempt to shed light upon the long-term stability of slave families in northern Virginia, lowcountry South Carolina, and southern Louisiana, respectively, this chapter examines the extent to which family bonds were forcibly ruptured, particularly by estate divisions, sale (local and long distance), and long-term hiring. In the following, it becomes clear that the boundaries inhibiting long-term stability among slave families, and the opportunities available to either avoid or soften the blow of forced separation, differed across time and space according to the nature of slave-based agriculture.

The Looming Threat: Fairfax County, Virginia

Most slave families in antebellum Fairfax County were forcibly separated from the outset. As shown in chapter 6, the dwindling size of local slaveholdings created a social landscape in which cross-plantation marriages among slaves were especially prevalent. With slave-based agriculture in decline, however, cross-plantation families found themselves highly at risk of becoming even more scattered, while co-residential families saw their relative stability regularly threatened by potential ruptures. In slave quarters throughout the county, enslaved people lived in constant anxiety of the very real possibility that they would one day be forcibly removed from family members. Opportunities to prevent it from happening were few, but when scattered locally, slaves often went to extraordinary lengths to maintain contact with their loved ones, employing familiar methods to transcend physical boundaries.

Estate divisions in particular confronted local enslaved people with formidable obstacles to maintaining stable domestic arrangements and social relationships. Because of the unusually small number of slaves living on most holdings in northern Virginia, severing at least some family ties during estate divisions was practically unavoidable. Families who lived on farms, as most slaves did, were particularly at risk. Joshua Buckley owned only eight slaves upon his death in 1821: three men, one woman named Fann, and her four children. With exactly eight heirs, however, Buckley's slaves were all separated from each other, including Fann's four small children from their mother. The three female-headed families owned by Peter Coulter likewise failed to escape forced separation. In his last will and testament, Coulter specified that his daughter Margaret receive "a negro woman call'd Mary & all her increase except a negro boy called William." His daughter Cordelia was to receive "one negro woman call'd Rachael & all her increase except Eliza, Winny & Joseph." And John Coulter, the only male heir, was to inherit "old Jinny, her son James, little Mary, Hannah & all her increase, also a negro boy called Joe." Joe was presumably Joseph, the above-mentioned son of Rachael. Finally, Coulter specified his wish that the remaining slaves, Henry, William, Eliza, and Winny—the latter three being the sons and daughters of Mary and Rachael, respectively—be sold in order to pay off his outstanding debts. Such cases appear in the will books with staggering frequency and appear even more frequently in the late antebellum period, when most slaveholdings had shrunk to minuscule proportions.[2]

In the early nineteenth century, manumission was perceived by a number of slaveholders as a viable and socially acceptable way of getting rid of surplus slaves during estate divisions, and a few local slaveholders followed their neighbor George Washington's example in freeing at least some of their "people" in their wills. Others devised arrangements for term slavery. In the 1820s, for example, Levy Stone commanded his heirs to "liberate and set free all my said slaves" as soon after his death as possible, as did Robert Gunnell. Ann Boggess expressed her desire to "set free, manumit and release from all servitude" her slave named Daffney; Daffney's daughter was to serve until the age of eighteen, her eldest son until he was twenty-one, and her youngest son, who was mentally handicapped, was to serve his mother (in bondage, curiously enough) for life. William Watters had already promised his young slaves their freedom while he was still alive: for the men when they turned twenty-five, and for the women when they turned twenty-one. In his will he specified that if any still be in bondage upon his death, however, they were immediately to be set free, probably so as not to encumber his heirs with the financial burdens of their care. By the 1850s manumissions were less common, but not unheard of. One P. J. Reid ordered his heirs to "take the Negroes to Penn[sylvani]a, where they will be Free."[3]

Ironically, however, manumission may have actually perpetuated forced separation in many cases, as Virginia law demanded the removal of manumitted slaves from the state. For those with family members still in bondage—especially cross-plantation families—freedom may have therefore been bittersweet. One young Fairfax County slave told reporter James Redpath that he would like to be free only if the law would allow him to stay in the area. "But it won't 'low me," he lamented. "I would have to go to Canada, or somewhere's else . . . [and] my mother has no other child but me." Many freed slaves took up residence in the District of Columbia in order to remain close by, hoping to one day save up enough money to buy their family members out of slavery. Few managed to actually do so.[4]

Slave families were not passive when faced with the prospect of forced separation through an estate division. Threatened with the establishment of physical boundaries between themselves and their family members, slaves often attempted to negotiate with the heirs of their masters to be bequeathed together, or at least with their youngest children. They were obviously not always (or even often) successful, but the few cases in which slaves were bequeathed along with family members may very well have been the result of such negotiations. One cross-plantation family managed not only to be united upon the deaths of their owners, but even to be emancipated in 1803. "Negro George," owned by one Compton, was bequeathed upon his master's death to his son, John Compton, with whom George probably grew up. When the master of George's broad wife and two sons died as well, George convinced his new master to buy his family; they were all subsequently granted their freedom. In 1836, a slave woman named Susan convinced her master to bequeath both herself and her daughter to her free husband, Benjamin Bronson, thereby uniting a divided family."[5]

Opportunities to prevent dispersal were rare and often ineffective, but when families were scattered locally, as they often were upon estate divisions, they softened the blow of separation by maintaining contact across property lines, visiting each other on the weekends. Some risked truancy to spend more time with their families. Harriet Williams, who absconded in 1855, was suspected of hiding out in Washington, where "she has relations." An advertisement for a runaway named Daniel Solomon hinted that he had extensive "relations in Fairfax County and the District." Although family members who were removed beyond a radius of ten to fifteen miles could not be visited with regularity, some slaves did procure permission from their masters to traverse great distances to see them. Nelly Shanks, an elderly slave at Bush Hill, was separated from her daughter Clarissa and her grandchildren in an estate division, the latter being sent to Farmington, some one hundred miles distant. In October 1827, Clarissa was fortunate enough to obtain special permission from her master to travel all the way to Fairfax County with her husband and three youngest children in order to see her mother. Such cases demonstrate that whenever possible slaves attempted to make the best of a bad situation by seizing opportunities to maintain family contact.[6]

With slave-based agriculture in decline, however, and the number of slaves that could efficiently be employed in mixed farming limited, it is perhaps unsurprising that the heirs of Fairfax County slaveholders did not always wish to hold on to their inherited bondspeople. Upon estate divisions, slave sales were the order of the day, as local residents quickly converted their inheritance into cash. This enabled them to pay off debts and spared them the financial burdens of having to maintain surplus slaves; in the process, however, they ruptured countless slaves' family ties.

Many such transactions were local affairs, with slaves being sold at local auctions to new masters in northern Virginia or Washington. Evidence of this can be found for the entire antebellum period. The eight slaves owned by Edward Blackburn were sold by his heirs to four different local buyers, two mothers being sold along with one infant each. A slave family owned by the heirs of Wormley Carter, consisting of an elderly woman, a man and his wife, and their two young children, were sold to three local buyers: the man and his wife were sold to one buyer, their two children to another, and the elderly woman to yet another. Such cases appear with frequency in the estate accounts of Fairfax County slaveholders. The *Alexandria Gazette* is also full of advertisements placed by "commissioners" and "trustees," announcing the local sales of slaves from the estates of deceased slaveholders. The following advertisement, which ran in the winter of 1851, was fairly typical: "FOR SALE . . . 9 likely NEGROES, of various ages, consisting of men, women and children, and who belong to the estate of Richard S. Windsor, late of Fairfax County, deceased." No provisions to keep family members together were specified.[7]

As stated above, Fairfax County slaves employed the pass system to maintain contact with family members who had been sold locally. Therefore, the result of local sales, like estate divisions, was often not the destruction of slaves' family ties but rather the destruction of their domestic arrangements. Co-residential families were turned into cross-plantation families, and cross-plantation families became even more scattered across the landscape. The real danger for the enslaved person in local slaveholders' eagerness to rid themselves of surplus slaves lay in the prospect of being sold to interstate traders and sent to the cotton South, where slaves could command higher prices but where they would never see their family members again. The most recent estimates by Steven Deyle indicate that between 1820 and 1860 some 875,000 slaves were forcibly removed from the Upper South, between 60 and 70 percent of whom were transported via the domestic slave trade. Along with this massive forced migration came unspeakable loss for slave families in the exporting states. Michael Tadman has calculated that the domestic slave trade probably destroyed one in three first slave marriages in the Upper South; the proportion of children separated from both parents may have been as high as one in five. Families in Fairfax County were particularly at risk because the lure of some of the South's most notorious interstate slave traders was so temptingly close by—Franklin & Armfield; Price, Birch & Co.; John W. Smith; E. P. Legg; and Joseph Bruin, among others, all operated from nearby Alexandria and Washington.[8]

Population data indicate that large numbers of slaves were forcibly removed from Fairfax County during the antebellum period. As stated in chapter 5, census returns show that the local slave population was reduced by no less than 47 percent between 1810 and 1860—not in relative numbers, but in absolute numbers. If projected population growth is taken into account, the decline was in fact much greater. In the decade preceding the Civil War, for example, the slave population declined by only 2 percent, but considering local mortality rates and birthrates during the 1850s, the real decline was probably more like 16 percent.[9]

It is important to consider that not all of these slaves were sold—a few were emancipated and many others were forced to migrate west with their masters, especially in the early decades of the nineteenth century. Between 1810 and 1830 several leading Fairfax County families departed for Kentucky, Ohio, and Missouri. Charles Love, one of the largest slaveholders in Fairfax, emigrated to Tennessee in 1820 "with his whole family," including his slaves. By the late antebellum period, emigration had passed its peak, but even in 1846 planter Barlow Mason departed "for Louisiana, together with his blacks, to settle there permanently." Westward migration may have spared local slave families from the humiliating auction block, but it also dissolved cross-plantation marriages and cross-plantation family relations. One enslaved man named George Williams, who was born in Fairfax County but as a small child emigrated with his master to a farm in Kentucky (and later fled to Canada), recalled that during the move west his mother was especially "grieved at leaving her husband," who lived on another farm.[10]

There can be no doubt, however, that the vast majority of slaves who left the county during the antebellum period did so in chains, sold to the burgeoning plantations of Georgia, Alabama, Mississippi, Louisiana, and Arkansas. Local traders and southern planters alike ran countless advertisements in the *Alexandria Gazette* promising distressed slaveholders cash for likely young hands under the age of thirty. Franklin & Armfield offered sellers in 1828 "Cash for one hundred likely YOUNG NEGROES of both sexes, between the ages of 8 and 25." Price, Birch & Co. wished "to purchase any number of NEGROES, of both sexes, for cash." And Joseph Bruin offered to "pay liberal prices" for "any number of NEGROES."[11]

Such ads caught the attention of the local slaveholding population. During estate divisions, many heirs of unwanted bondspeople immediately delivered them to interstate traders or sold them at auction to agents of southern planters. The five slaves owned by the estate of William Lane in 1829 were sold to slave trader Alexander Grigsby for a total of $1,287. In 1835 the five heirs of Francis Lightfoot Lee of Sully plantation instructed an agent to "see if it will be possible to get any or all of those negroes off. The sooner the arrangements are made, the better." And the heirs of Ann Mason, John Huntington, James Potter, and Sarah McInteer all sold slaves to local trader Joseph Bruin in the 1850s.[12]

Many local slaveholders, however, did not wait for their heirs to dispose of their surplus bondspeople. There is ample evidence that farmers and planters culled their labor forces in times of economic difficulty. One resident of Alexandria told a visitor in the 1830s that "if it were not for this detestable traffic, those who have a large number of slaves upon poor land (such is most of the soil near Washington), would not long be enabled to hold them; as it generally takes the whole produce of their labor to clothe and support them." Bushrod Washington of Mount Vernon sold fifty-five of his slaves to a Louisiana planter in 1821 for the sum of $10,000. As justification he claimed that he had struggled for twenty years to turn a profit from the "products of their own labor," but his slaves being "worse than useless," and his plantation losing between $500 and $1,000 per year, he felt he had no choice. The slaves who remained told British visitor E. S. Abdy "that the husbands had been torn from their wives and children; and that many relations were left behind." Washington's neighbor Lawrence Lewis, master of Woodlawn, was compelled to do the same. In 1837 Lewis wrote to an agent in Louisiana about the "prospect . . . of either selling [the slaves] or hireing them out; the loss I have met with will make me prefer the former rather than the latter." Richard Marshall Scott Jr., made a habit of selling surplus slaves whenever they misbehaved, probably so as to reconcile his financial troubles with any moral difficulties he may have had concerning sale. On 11 August 1846 he "sold my woman Catherine very much against my feelings, for $480 to Joseph Bruin," mainly as a result of her "improper conduct for some time past." By 1850, however, he had grown more callous. On 5 October he "sold our man Aaron today to B. O. Shekell for $200 on condition that said Shekell would remove him from the state." On several other occasions he delivered slaves to interstate traders in Alexandria and Baltimore without mentioning any misbehavior.[13]

Interstate traders sometimes roamed the countryside, visiting farms and plantations in search of potential sales. One northern Virginian wrote that "the dealer goes from house to house plucking the flower from every flock." Many sellers discreetly arranged for traders to pick up their slaves in the early hours before the sun rose, so as not to cause a scene. Benjamin, a slave from Fairfax County, testified that "frequently slaves would be snatched up, handcuffed and hurried off south on one of the night trains without an hour's notice." This was a common occurrence, and slaveholders exploited their slaves' anxiety by threatening to sell them if they did not work hard enough. One local bondsman told interviewers that "Master used to say, that if we didn't suit him, he would put us in his pocket quick—meaning he would sell us."[14]

Such was the desire to sell in the Potomac region that as early as the 1830s Franklin & Armfield, by far the largest and most successful slave-trading firm in the area, was reportedly sending more than a thousand slaves a year to the Deep South, with tragic consequences for slave families throughout northern Virginia and southern Maryland. Ethan Allen Andrews, a northerner charged with reporting on the domestic slave trade to the American Union for the Relief and Improvement of the Colored Race, visited the firm on Duke Street in Alexandria in 1836, and discovered that "in almost every case, family ties have been broken in the purchase of these slaves." Speaking with another trader, Andrews was told that "he never separates families [upon sale], but that in purchasing them he is often compelled to do so, for that 'his business is to purchase, and he must take such as are in the market!'" When Andrews asked whether they often bought wives without their husbands, the reply was: "Yes, very often; and frequently, too, they sell me the mother while they keep her children. . . . The prices are *monstrous high*, and that, in fact, is the very reason so many are willing to sell." By the 1850s James Redpath learned from a Fairfax County slave that the separation of families through long-distance sale was "as common as spring water runs." The slave added: "A darkey's worth a hundred dollars as soon as he kin holler. That's what the white folks say bout here."[15]

Enslaved people confronted with the reality of long-distance sale encountered few opportunities to prevent it, but many nevertheless risked desperate attempts to escape their lot. When slaveholder Elizabeth Tyler died, for example, one of her slaves named Henry fled immediately in order to escape sale to the Deep South. He was caught in Baltimore and sent back to Fairfax County, upon which he chopped off his own hand, apparently in order to render himself worthless for sale. Even this did not save Henry—he was sold anyway, presumably at a great loss. Cornelius Whatson, "said to be the property of a Mr. Nelson, a southern trader, but formerly the property of Mr. Chichester in Fairfax County," was apprehended in Alexandria after a similar escape attempt. Sharper, a Fairfax County field hand, fled to his wife's residence when he heard that he was going to be sold in 1831. He remained at large for seven weeks but was eventually caught and deported. A local slave named Randolph fled to the North after "three of his brothers [were] sold South," and Oscar Payne did likewise after "three brothers and one sister [were] sold South." Some slaves selected for sale sought the help of free family members and even abolitionists. In January of 1850, one literate young slave woman named Emily Russell wrote a desperate letter to her mother, begging her to do something because "I am in [Joseph] Bruin's jail, [along with] aunt Sally and all her children, and aunt Hagar and all her children, and grandmother is almost crazy." Emily's mother arranged for abolitionist William Harned to offer to buy her daughter from the well-known slave trader, but Bruin demanded the exorbitant sum of $1,800 because "she was the finest-looking woman in this country." Within a week Emily was deported in a coffle going south; she died en route in Georgia.[16]

Attempts to prevent deportation were of course the exception rather than the rule and were usually unsuccessful. Most slaves sold south had little choice but to accept their fate. Not only did they face the trauma of deportation, but their family members suffered enormously as well, as is evident from the morosely related testimony of one Fairfax County slave named Lydia Adams (who was herself eventually forced to move west with her master, but later fled to Canada). "One by one they sent four of my children away from me, and sent them to the South," Adams told interviewers. "And four of my grandchildren all to the South but one. My oldest son, Daniel—then Sarah—all gone." Her only solace was in believing that her master would be punished in the afterlife: "I am afraid the slaveholders will go to a bad place [when they die]. I don't think any slaveholder can get to the kingdom." Such sentiments no doubt resonated with enslaved people throughout the county.[17]

Slaveholders in Fairfax County had an alternative method of ridding themselves of surplus bondspeople without selling them to the cotton states—they hired them out annually, usually to merchants and business establishments in and around Alexandria and Washington, but also to local farmers who lacked the capital (or desire) to purchase slaves. This was especially popular in the latter half of the antebellum period. For Fairfax slaves who were hired out, as Donald Sweig aptly put it, "hiring must have seemed little better than outright sale." Their masters entered into agreements and signed contracts with third parties that lasted for fifty-one weeks, allowing them only the week between Christmas and New Year to spend at home with their families before being hired out again.[18]

There are several reasons that slaveholders often preferred to hire out their surplus slaves rather than sell them. Sweig suggested that many did so because they "were educated men, sensitive to abolitionist pressure, who had a strong revulsion against selling their slaves." This may have certainly been true for some, but the most respected and educated gentlemen of Fairfax County—including Bushrod Washington and Lawrence Lewis—sold slaves to the Deep South when the going got rough. Moreover, the sellers and hirers of slaves were often one and the same. Advertisements such as the following were a frequent sight in the *Alexandria Gazette*: "FOR SALE OR HIRE—A likely WOMAN 43 to 45 years of age." There were in fact a number of advantages to hiring out rather than selling. First, it provided slaveholders with a regular income from bondspeople who were otherwise superfluous or unproductive. Those who hired slaves were responsible for housing, feeding, and clothing them during fifty-one weeks of the year, which spared slaveowners the financial burdens of having to care for them. On top of that, slaves were often hired out for 10 to 20 percent of their market value—in other words, a slaveholder could often earn more money by hiring out a slave for five to ten years than he could if he sold the same slave outright. Many slaveholders hired out their slaves for a number of years and then sold them, increasing their profits exponentially. The system also brought with it certain social advantages. It allowed owners to rid themselves of surplus slaves yet without losing their status as slaveholders, and it allowed nonslaveholders to put on airs of the slaveholding class. According to historian Jonathan D. Martin, "hiring became a slaveholding cure-all" in struggling regions of the South such as northern Virginia.[19]

Slave hiring was extremely prevalent in antebellum Fairfax County, especially toward the end of the period. Frederic Bancroft calculated that during the year 1860 alone some 25 percent of the Fairfax slave population was forcibly hired out; Donald Sweig put the estimate around 28 percent. Estate divisions again seem to have been the most common circumstances under which slaves were hired out. The heirs of George Sweeney hired out his slaves for fourteen years straight, from 1834 through 1847. Matilda West's heirs likewise hired out her slaves for thirteen years, between 1809 and 1822. When Richard Marshall Scott, Sr., died in 1830, his slaves were hired out for sixteen years, until his young son, Richard Marshall Scott Jr., was old enough to take over the family estate. Sweig found that of all the Fairfax County estate accounts of deceased slaveholders between 1830 and 1860, some 35 percent mentioned slave hiring.[20]

The threat of forced hiring was not limited to estate divisions, however. In financial straits, local slaveholders routinely brought their surplus bondspeople to the county courthouse or other auction houses at the end of the year in order to hire them out to third parties, including the abovementioned Richard Marshall Scott Jr. On 1 January 1848 he jotted down a typical New Year's entry into his diary: "Attended at Samuel Catts for the purpose of hiring my servants. Hired 4 men, 6 women, 2 boys, 4 girls." He did this every year from 1846 till his death in 1856, and it provided him with a great deal of his income. Local newspaper announcements beckoned distressed farmers and prospective employers to meet at several locations throughout the county: "The days for hiring servants in Fairfax Co. are as follows—Fairfax Court House on Monday, December 30; Dranesville on Tuesday, December 31; Centreville on Wednesday, January 1, 1851." Visitors to Alexandria and Fairfax such as James Redpath, E. S. Abdy, Ethan Allen Andrews, Phillips Brooks, and Frederick Law Olmsted all remarked on the prevalence of hired slaves at local hotels and other establishments.[21]

In virtually all of these cases, enslaved people were forcibly separated from some or all of their family members for years at a time. The slaves attached to the John Fitzhugh estate were all hired out to different employers between 1816 and 1820, as were the bondspeople belonging to the estates of William Lane and Alfred Offret. The five slaves owned by William R. Newman, consisting of two women and their three children, were also hired out separately in the 1850s. Sometimes new mothers were hired out together with their youngest children, usually at a lower price, as the children were considered a burden.[22]

Local bondspeople were even sometimes removed from the region altogether to work for distant employers. In 1838 Winifred Ratcliffe hired her slave Moses out to a Mississippi employer for the exorbitant sum of $350 per year; she also sent a mother and child to Mississippi for $200 per year. Sending slaves farther south to work by the year was not only especially profitable but it also discouraged flight to the North, something many local slaveowners feared in the late antebellum period. After an escape attempt by a number of his newly inherited slaves at Arlington plantation, Robert E. Lee hired them out in southern Virginia, "where the border would appear less tempting." From such distances these bondspeople were denied the chance to see their family members even during the last week of the year.[23]

Enslaved people in Fairfax County often attempted to seize or create opportunities to gain some measure of control over the terms of their hire, supporting Jonathan Martin's argument that hired slaves resisted or attempted to profit from the system in a wide variety of ways. A few were permitted to keep a portion of their wages or earn wages for extra work, allowing them to purchase small gifts for family members or even put money aside in the (often delusional) hope of purchasing freedom. James Redpath heard from a local enslaved man that hired ferry workers on the Potomac were sometimes able to earn $6 a month for themselves. But the man hastened to add that many masters, including his own, did not allow their hired slaves to keep any of their earnings besides tips. Remarkably, some slaves tried to turn hiring into a vehicle for *uniting* nonresident family members by negotiating with their masters to be hired out to free relatives or the owners of their spouses. Henry Hubert, a free black man, was able to hire his enslaved sons from the estate of Harrison Allison between 1835 and 1840. One Letty, of Wilton Hill plantation, was hired out to Thomas Janney, the owner of her husband. At Bush Hill a woman named Ellen Ann, the wife of a free black man named David Grey, managed to get hired out to her husband in 1848, a year which she spent mostly pregnant and in which she was delivered of a child. This arrangement was probably permitted only because she was pregnant, however, because the following year she was hired out "to Murray Mason . . . not to husband," undoubtedly because the former could offer more money.[24]

More often, perhaps, enslaved people attempted to frustrate the deals made between their owners and new employers, pitting their two "masters" against each other. Many exploited their owners' fears of property damage and complained to them of physical abuse by their employers. In a diary entry from 1857, Virginia Gunnell Scott wrote: "Servant woman Margaret came home this day complaining of being badly used by Mr. Fairfax, manager for Mrs. Daingerfield," to whom she was hired. This strategy backfired, because Scott decided to handle the situation by simply selling Margaret to a local slave trader for $1,000, but others were more successful. Still other slaves made their seething anger so obvious to their new employers that the latter quickly found them all but unmanageable. James Frans-Tracy, an Irish newcomer to Fairfax County who depended on hired slave labor in the 1820s, described one hired woman named Anna as having "the most abuseive tongue, and wicked expressions that ever came out of any Person's mouth." He also had difficulty with a young slave girl named Hariot, whom he was "obliged to send . . . home, as I have tried every means in my power but can not make any thing of her. . . . She is the worst dispositioned child I ever knew."[25]

When all else failed, many slaves fled, sometimes successfully, sometimes not. Virginia Gunnell Scott became worried one day in 1859 when she heard her "servant man West, who was hired in Alex[andria], had left his place and had probably gone off in a northern vessel." But that same week West was caught when he showed up at his wife's residence "this morning at 3 o'clock," after which Scott "had him lodged in jail." George Coleman successfully fled Fairfax County after being very "unhappy" and continually "hired out amongst very hard white people." Even young slaves often ran away from their new employers, sometimes hiding out in the neighborhood of their parents, as was probably the case with one boy declared missing in 1853: "RAN AWAY—My boy BOB, aged about 12 years, absconded from the place where I had hired him." For enslaved people living in antebellum Fairfax County, hiring was just one of the many obstacles that continually threatened their domestic arrangements and family stability.[26]

A Relative Peace: Georgetown District, South Carolina

Slave families on the rice plantations of the lowcountry were of course also vulnerable to the prospect of forced separation, but the odds that any given family unit would actually be torn apart there remained consistently lower than in other parts of the South. Whereas in Fairfax County few families made it through slavery completely intact, most slave families in Georgetown District enjoyed relative peace and stability over time. The large size of local slaveholdings and the profitability of rice made the forced separation of slave families there unlikely, though not impossible. Even when faced with the threat of separation, however, local enslaved people attempted to seize or create opportunities to prevent it from happening or cushion its effects, often to some success.

During estate divisions, for example, slave families in the lowcountry were less often shuffled around or forcibly divided than their Fairfax County counterparts; in many cases they did not even change residence. The case of Henry Augustus Middleton is illustrative. Seeking a pardon after the Civil War, Middleton pleaded that while he had owned over 300 slaves in 1861, all of his slaves and their ancestors had been owned by the family since 1735, and that he had personally never bought or sold anybody, but had rather inherited them all with the plantation. This was not particularly unusual for Georgetown District. Indeed, local slaveholdings were remarkably stable from generation to generation, and visitors and residents alike frequently commented on the continuity of slave ownership within certain families. One former resident recalled long after the Civil War that most of the plantations in his neighborhood "had been in the possession of the same family for several generations, and the negroes had been born and bred upon them." J. Motte Alston claimed that "most of [my grandfather's] large number of negroes lived and died in the same family." Laurence Oliphant, who visited Georgetown District just before emancipation, noted that the slaves on one plantation "recall reminiscences of three or four generations back of the family to which they have belonged for nearly a century." And one visitor to the district remarked in 1836 that there were "many cases where successive generations of owners and slaves have run parallel to each other for a long course of years in the same families."[27]

Multiple plantation ownership among many local rice planters often made it possible to bequeath entire plantations with their slave populations largely intact, or at least greatly reduce the breakup of slave populations during estate settlements. Upon his death in 1827, for example, Plowden Weston bequeathed a plantation and over 250 slaves *each* to both of his sons. William Alston bequeathed to each of his three sons a complete plantation in 1838, with one of his sons even receiving two plantations. The hundreds of slaves attached to the estate were simply to be "divided equally" among his sons, "share and share alike." In 1853, Joshua John Ward likewise left his six plantations and more than a thousand slaves to be divided equally among his three sons, with each son receiving two plantations. So it was throughout the district, though not always on such a grand scale. John Hyrne Tucker willed his two plantations (Litchfield and Willbrook) to his two sons John and William Alexander in 1859—one counted ninety slaves and the other ninety-eight. Many slaveholdings in fact passed wholesale into the hands of planters' children even before the need arose for an estate settlement. J. Motte Alston recalled that at the turn of the century his "grandfather owned a number of plantations, many of which had been given away to his sons."[28]

Slaveholdings were moreover large enough that it was possible in most cases to keep families intact even when they did have to change residence. It is significant to note that in the inventories of Georgetown's grandees, slaves are frequently listed and appraised in simple family groups, unlike in northern Virginia, where they were usually listed singly and family ties were rarely mentioned. This suggests that slaveholders in Georgetown District in fact intended to keep families together during estate divisions. Doing so would have not only facilitated the equal division of such large groups of bondspeople but would also have prevented unrest among the slaves themselves during the settlement. Benjamin Alston Jr., for example, who owned 144 slaves when he died, only specified his wishes concerning eighteen domestic slaves in his will; the rest, numbering 126, were to be divided equally among his seven heirs, "share and share alike." In a subsequent inventory made in 1819 all but nineteen slaves were listed and appraised in collective family groups, indicating that they were to be bequeathed as such. In the slave inventories accompanying the will of Francis Withers, the owner of both Friendfield and Northampton estates, which together contained some 468 bondspeople in 1841, the slaves were also appraised collectively in family units. Likewise, the slaves belonging to J. H. Allston in 1846, as well as those belonging to the Gourdin family's plantations in Georgetown District, were bequeathed in family lots. Evidence from throughout the antebellum period suggests that this was the rule rather than the exception.[29]

When enslaved people in Georgetown District were transferred from one holding to another during estate divisions, thus, it did not often result in the destruction of simple family ties or marital arrangements. That is not to say that all slave families were protected from forced separation during estate divisions, however. First, even if simple families were often bequeathed intact, they were frequently separated from extended family and friends, including the adult sons and daughters of some couples, making them dependent on the pass system to maintain contact with those outside their immediate households. And second, even simple families were sometimes dissolved. Members of cross-plantation families in particular ran the risk of being removed beyond the immediate vicinity of loved ones, making regular visiting difficult or even impossible. Some slaveowners also bequeathed favorite (usually domestic) slaves individually to certain heirs. Nevertheless, most nuclear slave families in Georgetown District had less to fear from estate divisions than slave families in Fairfax County.[30]

Although many lowcountry rice planters were wealthy beyond the wildest dreams of most American slaveholders, certainly those in northern Virginia, slave sales were not uncommon in Georgetown District. Auctions were held with some regularity, especially during estate settlements or upon the retirement of some planters. Many such auctions consisted of small groups of slaves whose sale was intended to round out partitions or pay off any outstanding debts. The *Winyah Intelligencer* ran numerous advertisements for local estate auctions such as the following: "I will sell . . . 8 prime and very valuable NEGROES, the property of the Estate of the late Francis Brudot, to wit: Molly, Hester and Tenah her child, Hannah, Judy and George her infant, Silla, and Jem"; and "Estate Sale: On Saturday the 10th day of January next before the Court House will be sold FOUR NEGROES, belonging to the Estate of O. Potter, deceased." Some sales were much larger. The heirs of planter Thomas F. Goddard offered for sale "from FIFTY to SIXTY PRIME NEGROES" in 1832.[31]

Labor demands within the district were such, however, that a majority of slaves who were auctioned off at estate sales were eagerly snatched up by local planters, most of whom constantly kept their eyes open for bargains. Indeed, continuous purchasing by the most successful planters resulted in significant consolidation and growth of local slaveholdings during the antebellum period, as discussed in chapter 5. The account books belonging to planter Stephen D. Doar, for example, indicate that he was a regular frequenter of his neighbors' estate sales. In 1855 he purchased at least forty-four slaves at three local estate sales. Robert Allston continuously expanded his labor force through local purchases during his long career. In 1828, for example, he bought thirty-two slaves from the estate of William and Sarah Allen. In the 1850s he purchased forty-two bondsmen from Thomas Pinckney Alston (whose sons did not want to take over the family business when he retired, inducing him to sell 177 of his slaves). In 1857 he bought another group of fifty-three Waccamaw slaves and two years later he purchased Pipe Down plantation in its entirety, complete with 109 slaves, from the widow Mary Ann Petigru, who wished to cash in on her estate when her husband died. Finally, at the auction of one Mrs. Withers in 1859, Allston bought an additional forty-one bondspeople for his son Benjamin.[32]

Slaves who were sold locally attempted to maintain family contacts by requesting passes for weekend visiting, as was the case in northern Virginia. But a handful of runaway slave advertisements illustrate that bondspeople who had been removed beyond a reasonable distance sometimes risked truancy to visit with family and friends left behind, not unlike their counterparts in Fairfax. One missing man named Adam had been bought "of the Estate of John Bowman, of South Santee," and was presumed to have returned to the Santee area to visit family and friends. Another missing slave who absconded with his wife and four children was suspected of being harbored by extended family or friends near "the plantation of Thomas W. Price Esq, at the sale of whose negroes he was purchased about two years ago." Sammy, who was owned by a planter on the Sampit River, had "a mother at Rose Hill (Waccamaw) where the subscriber understands he is."[33]

Such incidents appear to have occurred with less frequency in Georgetown District than they did in northern Virginia, however, probably because families were less often broken up through sale. As was the case during estate divisions, slave families in the lowcountry were usually sold in family groups. In the records of Paul D. Weston, for example, an inventory of 145 slaves "to be sold" listed and appraised them all in family groups. Likewise, the slaves belonging to the estate of E. J. Heriot were all sold in family groups. Sale advertisements often announced slaves in family units. The auction house of John Shackleford & Son in Georgetown advertised in 1819 the sale of "Nine valuable Negroes," adding that "these negroes being one family, cannot be separated." The fifty-three Waccamaw slaves who were snatched up by Robert Allston at an auction in 1857 were also advertised and sold in families, and Richfield plantation was advertised for sale along with "a gang of about 111 Negroes . . . to be sold in Families." Purchasers advertised to buy slaves in family groups as well. One planter wanted "a family of 6 or 8 Negroes." Another promised "cash . . . for a family of 8 to 10 healthy negroes."[34]

Purchasers of slaves had good reasons to want to buy in family groups; slaves were cheaper when purchased in lots rather than individually. Robert Allston advised his nephew: "If you meet with an orderly gang of some planter, neighbor or otherwise, who would sell you at $500 [per slave] rather than send away and separate his people at a higher figure then in such a case you might adventure." Slaves were not only cheaper when purchased in family groups, however, but also more efficient, making the actions of the enslaved an important factor in maintaining family stability over time. Enslaved people throughout the lowcountry indeed created opportunities to keep simple families intact by making it unattractive to acquire them otherwise. As was the case when they were confronted with what they deemed to be unreasonable tasks, bondspeople in Georgetown District protested the separation of family members during sales and estate divisions. Historian Leslie Schwalm found that in the lowcountry "the pending sale and separation of members of a slave community created a 'general gloom' that settled on the plantation slaves 'at the idea of parting with each other.'" Their reaction had a negative effect on plantation labor, and indeed "so disrupted the peace and efficiency of the slave workforce that it became common wisdom among nineteenth-century rice planters that slaves should be purchased and sold in intact family groups." Doing otherwise could have undesirable consequences. According to one planter, if slaves were bought individually and thrown "all together among strangers, they don't assimilate, & they ponder over former ties, of family, &c., & all goes wrong with them." Adjusting to the demands of the market, auctioneers and sellers of slaves offered them in family groups.[35]

Not only did enslaved people seize opportunities to keep their family ties intact during the transfer of property, whether by estate divisions or sale, but on rare occasions they even attempted to keep entire plantation communities intact, sometimes successfully. Take for instance Robert Allston's purchase of Pipe Down plantation in 1857, complete with its 109 slaves. According to his daughter Elizabeth Allston Pringle, the purchase was largely the result of the slaves' actions. "As soon as Uncle Tom died," she remembered, "Aunt Ann wrote to my father, asking him as a great favor to buy her plantation and negroes," but her "father replied immediately that it was impossible." However, "then the negroes from Pipe Down began to send deputations over to beg my father to buy them . . . saying they had fixed upon him as the owner they desired. At last my father conceded." By convincing one of the wealthiest planters in the district to purchase them, the Pipe Down slaves were spared the auction block and the breakup of their community. They were also assured relative stability in the future. Thirty-six were transferred across the river to Guendalos, but presumably in family groups, as was customary.[36]

Neither did the threat of long-distance sale loom over slave families in Georgetown District to the extent that it did in northern Virginia. Georgetown planters were simply not as hard-pressed to rid themselves of their bondspeople. Indeed, most constantly sought to expand their slaveholdings, as stated in chapter 5. Unlike the Fairfax County slave population, which plummeted by 47 percent between 1810 and 1860, the Georgetown slave population increased by 53 percent during the antebellum period, despite far higher mortality rates and lower birth rates. Only during the 1830s did it slightly diminish, which may have been the result of either increased mortality or emigration and sale among the smaller planters who could not compete with the grandees. If projected population growth is taken into account for the 1850s, the local slave population grew less than expected by only 160, suggesting that perhaps that number of slaves were deported during the decade—but in a district containing over 18,000 bondspeople, even this was not a significant proportion of the slave population.[37]

Despite the fact that South Carolina as a whole was a net exporter of slaves, evidence suggests that the rice planters of Georgetown District were more prone to import bondspeople from other parts of the state or South than they were to export them to the cotton regions. In the local periodical *Winyah Intelligencer*, for example, advertisements for fresh arrivals such as the following were a frequent sight: "Negroes for Sale. Will be sold . . . on the 10th of February next, 50 to 100 likely NEGROES, in families. [From] Wilmington, NC." Such advertisements were absent in the newspapers of northern Virginia. Interviews with former slaves also provide evidence that local planters were primarily the purchasers rather than the sellers of the antebellum slave trade. Welcome Beese, owned by Joshua John Ward, related that his master "bought my mother from Virginia. Dolly [was her name]." Lizzie Davis's grandmother was sold to Georgetown District from Marion District, South Carolina: "I remember, I hear my father tell bout dat his mammy was sold right here to dis courthouse [Marion], on dat big public square up dere, en say dat de man set her up in de wagon en took her to Georgetown wid him. Sold her right dere on de block. . . . Pa say, when he see dem carry mammy off from dere, it make his heart swell in his breast." Historian William Dusinberre found that "a host of new slaves were imported into [the] Waccamaw plantations" in the decades before the Civil War.[38]

Certainly not all local slaves were spared deportation during the antebellum period, and it is important not to exaggerate their protection from the interstate trade. Newspaper advertisements such as the following, placed by a slave trader, were a rare sight in the district, but they did occasionally appear: "Negroes Wanted. Persons who wish to sell NEGRO MEN from 8 to 25 years of age, can have CASH to them." Research by Michael Tadman has revealed that Georgetown District counted two registered slave traders in the nineteenth century: G. S. S. Christie, who operated from 1845 to 1857; and S. Coletrane, who traded in slaves during the 1850s. Considering the large number of slaves resident in the district, however, having only two such registered traders, who moreover only operated for a relatively short period of time, suggests that there was no booming slave market in the Winyah Bay area. Indeed, it is possible that these traders imported slaves rather than exported them.[39]

But local planters could always sell to interstate traders based in Charleston or elsewhere, and many did with few qualms about forever separating their bondspeople from family and friends. When the above-mentioned Thomas Pinckney Alston retired to Georgia in 1859, he decided to sell 177 of his slaves "at the best attainable price." Forty-two were bought by Robert Allston, but the rest were "scattered to the wind." Robert Allston himself was forced to sell fifty-nine slaves when he inherited his brother's Waverly plantation in 1834, which was encumbered with a heavy debt of over $50,000. Luckily for the slaves, their mistress's own brother—who lived in Charleston but was the absentee owner of a cotton plantation in Georgia— agreed to buy fifty-one of them in family groups. Even then, parting with extended family and friends was traumatic enough that Allston decided to give them only twenty-four hours' notice, presumably to prevent negotiation attempts, protest, or flight. Upon arrival in Charleston, from where they were to be deported, the driver (who had been selected to remain but asked to accompany the deportees) nonetheless sought out his mistress and successfully convinced her to keep his grown daughter Sarah and her partner Bob. The rest were sent to Georgia. Otherwise Allston never sold any slaves, save one whom he sold to New Orleans in 1847 for "gross and wicked misconduct." Plowden Weston also sold over a hundred of his slaves from his plantation Laurel Hill, which he decided to get rid of in 1856 in order to concentrate his holdings around Hagley, his resident plantation. The purchaser of the plantation, Daniel Jordan, offered to buy the slaves as well, but Weston decided instead to sell most of them to the Deep South where they could command a higher price. Their departure by riverboat was recorded by Weston's wife Emily: "Tears filled my eyes as I *looked* and *listened* to the wail from those on shore echoed by those on board." Her husband was less perturbed.[40]

Most slave families in antebellum Georgetown District were spared such a fate, however. Unlike Thomas Pinckney Alston and Plowden Weston, many local planters sold their plantations with their slave populations intact, as was the case with the above-mentioned sale of Pipe Down to Robert Allston. Slave families in the district were also largely safeguarded from separation through long-term hiring, unlike families in Fairfax County. Few plantations in the lowcountry counted "surplus" slaves among their labor forces. Given the profitability of rice, its economies of scale, and the high mortality rates in the region, it was logical for planters to keep valuable field hands for themselves rather than hire them out to third parties. Indeed, many local rice planters hired slaves from elsewhere as temporary additions to the labor force. In December 1821, for example, the widowed Charlotte A. Allston hired on sixteen slaves for the 1822 planting season. One local planter advertised in 1829 that he "WANTED TO HIRE—From 12 to 15 Field Negroes, accustomed to the culture of Rice."[41]

That is not to say that local planters never hired out their slaves, but they did not often do so for long periods of time and frequently even kept simple family ties intact when they did. Robert Allston once hired out a number of his enslaved men to help construct a railroad—from which he stood to profit as the producer of a valuable market commodity and as a stockholder. This was a temporary project, however, and did not separate these men from their families for the entire year. James Ritchie Sparkman hired out five carpenters to work for William King in 1858, but only for one month. William Lowndes, on the other hand, annually hired out a number of his slaves to his brother in the early nineteenth century. In 1807 he mentioned in his account book that twelve of his bondspeople had "been working for some years with Mr. Charles Lowndes." However, these slaves were hired out collectively in three family groups. Again the threat of inefficiency seems to have convinced local slaveholders to keep families intact, a practice that appears to have been common. An advertisement for slave hire in 1825 offered "between twenty and thirty Field Hands, with their families."[42]

As was the case in northern Virginia, enslaved people in Georgetown District were most likely to be hired out during estate settlements, as is evident from the advertisements placed in the local *Winyaw Intelligencer*. Thomas Wilson's heirs offered for hire "A GANG OF NEGROES, Men, women and children, the property of Thomas Wilson's children." In 1830 the "SIXTY and SEVENTY NEGROES of the Estate of N. Snow, a lunatic" were offered for hire for a period of one year. Most lots offered for hire could hardly be called gangs, however. One group of slaves to "be hired on the 1st and 10th of January, before the Market in Georgetown" consisted of "FOUR NEGROES, the property of D. G. Williams, late of Georgetown, deceased." Another consisted of "SIX NEGROES, belonging to the Estate of the late Wm. M. Willson." It can be assumed that most such lots consisted of simple families. In all, only a small minority of slaves in antebellum Georgetown District were forcibly separated from family members through hire, sale, or estate divisions.[43]

A Volatile Society: St. James Parish, Louisiana

The extent to which slave families in southern Louisiana were disrupted by estate divisions, sale or long-term hiring defies easy categorization. While the profitability of sugarcane and the continuous growth in slaveholding size safeguarded many families from dissolution, St. James Parish remained a volatile slave society during most of the antebellum period. Overcapitalization, starting planters' heavy dependence on creditors, or a few bad crops in a row could all spell disaster for the sugar masters, inducing them (or their heirs) to rid themselves of some or all of their bondspeople. Slave families' experiences with forced separation in the sugar country therefore seem to occupy a middle position between those of families in northern Virginia, who were regularly torn apart, and families in lowcountry South Carolina, who were rarely dissolved. In St. James Parish it all depended, and enslaved people encountered few opportunities to gain any measure of control over their fate.[44]

Estate divisions in the parish appear to have combined characteristics found in both northern Virginia and lowcountry South Carolina. When they were *not* sold upon the death of their masters, slaves in St. James were often bequeathed to several heirs. And although slaveholdings were relatively large compared to holdings in other regions—especially in the late antebellum period—they were not as large as holdings in the lowcountry, which would suggest that family ruptures in southern Louisiana may have been a legitimate threat for many, as they were in Fairfax County. Unlike in Georgetown District, moreover, multiple plantation ownership in St. James was far from widespread, so bequeathing entire plantations intact to individual heirs was usually impossible. Historian Ann Patton Malone found that in Louisiana as a whole, estate divisions were a major cause for family disruption.[45]

Yet, intriguingly, the probate records for St. James Parish provide remarkably few examples of slaves actually being divided and changing residence during estate divisions. Plantations were more frequently kept intact and effectively turned into partnerships, jointly owned and operated by two or more heirs, either indefinitely or until one heir bought out the rest. In other words, the sons and daughters of sugar planters were more likely to inherit an *interest* in the slave population of a plantation than any specific group of slaves. This would have been an economically sound practice, considering the exorbitant maintenance and operation costs involved in sugar planting, and the profitability of economies of scale. What good would it have been for each heir to receive ten slaves? How would such narrow plantations be divided? Who would get the costly sugar house with its expensive machinery? As stated in chapter 5, the trend in the sugar country was toward growth and consolidation. If the heirs wished to extract maximum profit from their families' enterprises, they could better pool their resources rather than all becoming small cane farmers.

The succession papers of Leonard Fabre, for example, filed in March of 1828, mention that he owned one-fourth "of a certain plantation, situated in this parish . . . with the slaves, cattle, implements of husbandry and other appurtenances thereunto belonging." The plantation, called Fabre & Fabre and containing fifty-three slaves, had been inherited and was held and operated "in partnership with Marie Elizabeth Croizet [his sister], the widow Joseph Laurent Fabre [his deceased brother's wife], and Joseph Paul Fabre [his brother]." Upon Leonard's death his minor son inherited his father's quarter interest in the plantation and its slaves. Even into the third generation of ownership, thus, the slaves of Fabre & Fabre remained living on one residence, sparing families forced separation. The succession papers of Victorin Roman, which were first filed in 1832, also provide evidence of joint operations among several heirs. When Roman died in 1832, his estate was "greatly burthened with heavy debts and charges." Yet the family decided to continue operations as normal, dividing the proceeds equally.

When Roman's widow passed away in 1838 the children, led by the oldest son Jean Jacques, collectively took over, investing even more capital in the plantation. At a family meeting it was decided that Jean Jacques be empowered "to borrow, either from Capitalists or Banks . . . sums sufficient to satisfy the debts of said deceased and to buy an additional number of slaves to increase income of said property." During all of this the Roman slaves did not change residence, save one woman who was sold with her two minor children in 1838, perhaps for misbehavior.[46]

This practice appears to have been widespread among both large and small slaveholders throughout the antebellum period. In 1855 the five heirs of C. M. Shepherd—who, in community with his wife, held a one-fourth interest in Golden Grove plantation (containing 242 slaves)—received an almost comical "undivided one-fifth of the said undivided one-fourth of the Golden Grove plantation and the slaves thereto attached." Persac's map of St. James sugar plantations from 1858 shows several such joint ventures, including plantations owned and run by, among others, the Armant Brothers, Widow Priestley & Heirs, Estate of J. T. Roman, Widow J. Goutraux & Son, Bertaua Brothers, Mrs. Melancon & Son, and Mrs. G. Mather and Son. Slave families who were fortunate enough to live on such plantations were spared the fate of many of their northern Virginia counterparts, but there is little evidence to suggest that circumstances were greatly influenced by the slaves' own negotiations, as was clearly the case in lowcountry South Carolina. In the sugar country, many heirs simply calculated that they could make more money by keeping the family business—including its slave population—intact.[47]

Certainly not all were so inclined, however, for local estates were more frequently liquidated upon the death of a slaveholder, posing a far greater threat to slave family stability. The production of sugarcane was profitable indeed, but the dependence of starting planters on borrowed capital often burdened estates with debts of such magnitude that the widows and heirs of many sugar masters—especially young planters—quickly found themselves overwhelmed; more than a few decided against joint ownership and simply threw in the towel. This is clear from the probate records, which—like those of Fairfax County—are full of evidence that slave sales (and estate sales) were anything but uncommon during estate divisions in St. James Parish. The widow of planter Auguste Gaudet petitioned the parish court in 1832 to sell the plantation and all of the slaves belonging to her late husband's estate ("except the slave Marguerite and her issue") to "pay the debts left by him." The widow of Jean Baptiste Chastant also found her husband's estate "burthened with debts to a considerable amount," making it "absolutely... necessary to sell the greater part (if not the whole)," including most of the slaves and the plantation itself. Numerous other examples can be found in the succession papers, spanning the entire antebellum period.[48]

Some plantations were sold along with their resident slaves, as was often the case in Georgetown District, which would have prevented the disruption of enslaved people's domestic arrangements. In 1817 the brothers J. H. and R. D. Shepherd bought Golden Grove plantation from François Guerin together with all of its machinery and 116 slaves. John Burnside, the grandee who in 1857 purchased an enormous estate that straddled the border between St. James and Ascension parishes, likewise appears to have bought the plantation's slaves along with the land. Just four years later the British correspondent William Howard Russell was told by one resident slave that he had been born and "raised on the plantation," despite several changes in ownership. Newspaper advertisements sometimes specifically mentioned that plantations and slaves would be sold together. In 1849 the sugar farm and fifteen slaves belonging to the estate of Eboy Eber were put up for sale with instructions that they "shall all be sold in one lot and payable cash." Another plantation was offered for sale "together [with] 86 slaves," and yet another was offered for sale with specific instructions that "the immoveables and slaves . . . are to be sold together." Former bondspeople recalled such joint sales in their interviews by workers of the Federal Writers' Project. One woman named Melinda related that her grandmother had always lived on the same plantation, despite "the passin' of three generations." According to her, "the plantation had changed hands, but Grandma's lot remained untouched." Albert Patterson remembered that "our plantation was sold twice before de war," adding that both times "one man buy it all."[49]

A perusal of the probate records, however, indicates that the slave populations of many plantations were divided and sold individually or in small lots to several buyers. Many lots may have consisted of complete families. Travelers Francis and Theresa Pulszky were told by the daughter of one local sugar planter that "a good master never sells the husband from the wife," but unlike in Georgetown District, estate inventories in St. James rarely list slaves in family groups, save mothers with their minor children. They were thus probably not intended to be sold intact but rather in the most convenient manner possible. One local newspaper advertisement announced the sale of "fifty-one slaves, to be sold separately or by families," leaving it up to the buyer to decide whether to respect familial connections. The slaves belonging to Oak Alley plantation, owned by the estate of Jacques Telesphore Roman in 1848, were sold in lots and singly. The Roman heirs decided to sell only thirty-eight slaves in bulk with the plantation itself—the slaves and the plantation still being mortgaged to the Citizens Bank—but the other resident bondspeople were less fortunate. Twenty-four slaves were singled out in 1853 to be sold "singly and severally and . . . one by one," for example.[50]

The labor demand in the parish was so high that the frequenters of such sales were usually local planters or family members of the deceased, as was the case in the lowcountry, which suggests that if enslaved people were sold separately from their family members they probably remained within visiting distance. Twenty slaves were sold at the estate sale of planter Joseph Dugas in 1833 to thirteen different buyers, all of them neighbors and extended members of the Dugas family. Because the slaves' family ties were not recorded, it is unclear to what extent families' domestic arrangements were ruptured; ten of these slaves were sold as singles, but they may have been solitaries. Patrick Uriell's twenty-eight slaves were sold to three local buyers, four mothers being sold together with their minor children. Valcour Aime bought a number of slaves from his brother's estate auctions in 1840 and 1841, also including some mothers with their small children.[51]

The collective sale of mothers with their minor children was common because Louisiana forbade the sale of children up to ten years of age from their mothers (unless they were orphans), and the law appears to have been generally respected. When the heirs of Jean Baptiste Keller sold the slaves attached to the estate in 1833, for example, they demanded a down payment of "fifty dollars on each of them, not reckoning the children under ten years of age who may be sold along with their mothers." Despite the law, however, an aversion among slaveholders to separating family members does not appear to have been widespread. Orphan siblings were casually separated from each other. Two slaves sold at the estate auction of Zenon Arcenaux consisted of orphan sisters (ages four and seven) who were sold to different purchasers. Children over the age of ten were often sold apart from their parents, too, as is clear from the estate sale accounts of planter Pierre Chenet. Chenet's thirty-five slaves were sold to thirteen different buyers in 1837, all of whom were locals and two of whom belonged to the Chenet family. Four mothers were sold along with their minor children, but children over the age of ten were sold individually. Twelve-year-old Delphine, for example, was the only slave sold to one Louis Pollet, obviously separating her from her family.[52]

Although evidence is scarce, it seems likely that some enslaved people in St. James Parish at least attempted to create opportunities to avoid family separation by local sale—for example by negotiating with either the heirs of the estate not to be sold or with potential buyers to be purchased along with certain family members, as was common throughout the South. The widow of Auguste Gaudet originally stipulated to the parish court that all of the slaves belonging to her husband's estate be sold except one Marguerite, but later changed her mind and requested to keep "the slave Alexandrine and her children, Constant (5 years) and Theodore (3 years old)" as well, perhaps at the specific request of either Alexandrine or Marguerite. Some slaves attempted to take matters into their own hands: in June of 1854 a man and wife absconded from the plantation belonging to the estate of François Dufresne to avoid sale and separation. Yet attempts to prevent forced separation were obviously not always successful. Former bondsman Hunton Love recalled the sale of a group of slaves on his plantation: "Susan was bought and told to follow her new master. She was just about in childbirth and wouldn't move. When urged, [she] said, 'I won't go! I won't go! I won't!' For that she was given one hundred fifty lashes." When enslaved people were sold apart from family members, but remained in the general vicinity, they retained family contact by requesting passes for weekend visiting. As elsewhere, however, slaves sold beyond a distance of ten to fifteen miles were harder put to visit their families regularly, although even then some risked truancy to do so. One ex-slave recalled in an interview with workers of the Federal Writers' Project that his "father was sold up in Ascension Parish to a bad man, but he wouldn't work. So de man learned him a trade, and then he ran away to come back and see his family."[53]

Enslaved people appear to have been vulnerable to sale primarily during estate divisions, when many heavily capitalized estates were liquidated. The local nature of such sales suggests that family units were probably not often dissolved but that their domestic arrangements were disrupted. While their masters were alive, however, slaves enjoyed a relative degree of stability in an otherwise extremely profitable and wealthy region of the South, similar to the experiences of families in the lowcountry. The continuous growth in plantation size during the antebellum period, discussed in chapter 5, indicates that most plantations in St. James Parish were not broken up, and the explosive growth of the slave population (by 363 percent between 1810 and 1860) suggests that for slaves the chances of being sold outside of the parish or state were slim.

Long-distance sale always remained a possibility, of course, especially under certain circumstances. Historian Ann Patton Malone found that most Louisiana slaveowners "did not usually sell workers unless they were unproductive, considered troublemakers or confirmed runaways, or in the case of a serious financial reversal." In the first three cases bondspeople could avoid deportation by working hard and staying out of trouble, but in the case of financial reversal there was little room for negotiation. The widow of Patrick Uriell, whose late husband's estate was heavily burdened with debts, originally stated to St. James Parish officials that she believed "it would be more advantageous to the succession that they [the slaves] be sold at the City of New Orleans, by an auction," presumably because they could command a higher price there. For unknown reasons the slaves were eventually sold locally, as stated above, but the incident makes clear that local slaveholders had no qualms about removing their inherited bondspeople from the parish via the booming slave markets of New Orleans if they thought it convenient to do so. In practice, however, few enslaved people in the sugar country were deported. Frederick Law Olmsted spoke to a local slave, originally from Virginia, who claimed that "folks didn't very often sell their servants here, as they did in Virginia. They were selling their servants in Virginia all the time; but, here, they did not very often sell them, except they run away."[54]

Indeed, most of the evidence that pertains to the slave trade in the southern Louisiana sugar country deals with the *importation* of slaves from other parts of the South, especially the Upper South. Louisiana slaves found themselves on the receiving end—theirs was the final destination for the thousands who poured in through the South's largest slave market at New Orleans. Population growth in the parish was explosive, for example, despite having higher mortality rates than birth rates. Probate records for St. James Parish are also full of inventories listing "American" slaves, designating bondspeople who had been imported from outside of Louisiana. Of the twenty-four slaves belonging to Jean Baptiste Boucry in 1833, for example, thirteen were "American" and two older women were even African. Former slaves recalled not the departures of family members, as in northern Virginia, but rather the arrival of new laborers. Elizabeth Ross Hite for one told interviewers that she "ain't never seen an auction block, but I'se seen slaves when dey come off de auction block. Dey would be sweatin' and lookin' sick." Advertisements in local newspapers called not for able-bodied slaves to be deported elsewhere but announced the sale of new arrivals, such as the following: "NEGRES! NEGRES! Un grand choix d'esclaves des deux sexes" and "NEGROES FOR SALE. About one hundred Virginia slaves." Like the rest of the sugar region, St. James Parish was a net importer of slaves, not a net exporter—a factor that safeguarded many local slave families from potential dissolution.[55]

The insatiable demand for labor that safeguarded many enslaved people from being sold outside of the parish also protected most from being annually hired out to third parties. As was the case in Georgetown District, sugar planters did not often hire out their slaves—indeed, they regularly augmented their own labor forces with extra hands from elsewhere during labor-intensive periods such as the grinding season. Only smallholders seem to have occasionally hired out a few hands to their wealthier neighbors, and then only for short periods of time. The account books of Wilton plantation, owned by W. W. Wilkins, illustrate the trouble to which sugar planters went to secure extra labor when they needed it. Wilkins owned both a sugar plantation in St. James Parish and a cotton plantation in the northern part of the state, and in 1847 he diverted a number of hands from his cotton plantation to help his gangs in St. James during the grinding season. He also hired a few slaves from his neighbor and small slaveholder Octave Colomb. In 1853 Wilkins paid smallholders P. Webre and Andrew Crane $37.50 for slaves hired to do two days' windrowing, and in 1854 he paid M. Gourdin $70.67 for the hire of a number of slaves to perform three days' work. This practice appears to have been common throughout the sugar country, especially during the harvest. Samuel Fagot spent almost $850 on hiring slaves for the 1854 grinding season; Jean Baptiste Fernand's expenses for "loyer de negres" in 1858 amounted to $150. John Burnside even occasionally hired Irish free laborers to perform unhealthy jobs such as cleaning out ditches. His overseer claimed that it was "much cheaper to have the Irish do it, who cost nothing to the planter, if they died, than to use up good field hands in such severe employment."[56]

Slaves in the sugar country were usually hired out only for temporary stints when they were hired out at all, which did not really count as forced separation from family members. Even so, however, many enslaved people attempted to create opportunities to negotiate the terms of their hire, like their counterparts in Fairfax County. Most commonly they protested against being hired to planters whom they deemed to be inadequate or cruel. Andrew Crane, a starting planter who was occasionally dependent on hired labor for his cane harvests during the 1850s, had a particularly bad reputation among the local slave population. When Crane tried to hire a girl named Emma from E. Herbert for the 1858 grinding season, Emma—who had worked for Crane before—disrupted the negotiations by complaining to her master. Herbert replied in a letter to Crane that his slave girl "complains mightily about your feeding; if it is so my girl shall not pass the grinding season at your house. As for the rest of my boys, I have already hired them at Mrs. Heidi Nichols. I was to hire them to you, but they told me they would go anywhere before they would to you. . . . I hired them according to their wishes." Most hired hands were probably not as successful as Herbert's slaves in affecting the nature of their hire, but the case illustrates that neither did they always resign to their lot. A vast majority of enslaved families in St. James Parish were spared forced separation by long-term hire, however.[57]

Conclusion

The stability of enslaved people's family structures and domestic arrangements over time differed across space and depended heavily on both the nature of slaveholding and the labor markets of different slave societies. In Fairfax County small slaveholdings and the decline in regional agriculture proved disastrous for slave families, as estate divisions and long-term hiring severed domestic arrangements, and the domestic slave trade ripped many families completely apart. The nature of forced separation in Georgetown District was very different: large slaveholdings and multiple plantation ownership, as well as the profitability of rice cultivation, protected most slave families from being divided, sold, or annually hired out. St. James Parish again takes a middle position. Joint partnerships cushioned the effects of estate divisions, and both the profitability of sugar cultivation and insatiable demand for labor in the region safeguarded many families from sale and long-term hiring; yet the dependence of sugar planters on borrowed capital forced many heirs of deceased slaveholders to sell their plantations and slaves, sometimes as collective units but perhaps more often individually or in small lots. Louisiana law protected the separation of mothers from their minor children, but older children and spouses could be (and were) sold apart from their families.

In none of the three regions were slaves passive when confronted with the threat of forced separation, even if they were not always (or even often) successful. They negotiated to be sold or hired together, they ran away, and in lowcountry South Carolina they even made family separation an economically unattractive proposition for potential buyers or employers.

The boundaries and opportunities with which families in bondage were confronted to avoid separation and maintain a degree of stability varied from region to region, but they appear to have been most conducive to family stability in Georgetown District, and least so in Fairfax County.

IV

Conclusions

8

Weathering Different Storms

The secession of Virginia, South Carolina, and Louisiana from the Union in the winter of 1860-61 placed the institution of slavery in each of those states—as in the rest of the South—on what would ultimately prove the path to destruction. By the time the Civil War ended in 1865, the storm of bondage was over for more than four million African Africans, and many of the most formidable boundaries that enslaved people had encountered in shaping their family lives—from the pass system to the auction block—were shattered. Yet when the experiences of individual slave communities are illuminated, it becomes clear that more than one storm passed when the guns fell silent. Freedmen in Fairfax County, Georgetown District, and St. James Parish had survived very different ordeals. They had all been slaves, to be sure, but the nature of slavery in their respective regions had differed markedly.[1]

Family was the cornerstone of American slave communities and slave culture, but as this book has demonstrated, the specific nature of slave family life was contingent upon time and place, not only because the institution of slavery itself was susceptible to variation, but also because the extent and nature of agency among enslaved people varied as well. Certainly many factors influenced the nature of slave family life, but the dynamic relationship between the nature of regional agriculture on the one hand, and the boundaries and opportunities for family life on the other, was of particular importance in laying the foundations for slave families' daily experiences, domestic arrangements, and long-term stability.

This crucial link between different local economies and slave family life has long been underestimated in the historical literature. Past studies have tended to paint one-dimensional pictures of American slave families by underestimating regional economic differences and by ignoring or—far more often—overemphasizing the agency of slaves in shaping their own lives. It is in this context that the comparative approach is so valuable, as it offers a means for understanding slave families in different settings as they truly were: dynamic social units that were formed and existed under different circumstances across time and space.

As stated in the introduction, the three regions analyzed in this book represent the extremes of the southern economy. If the cotton districts were the "norm" in the antebellum South, the wheat, rice, and sugar districts were patently not. But it is particularly insightful to consider what life was like for slaves living in the non-cotton South, in regions that would be considered abnormal with respect to economic trends, working conditions, slaveholding size, and the supply and demand for slave labor. As historian Peter J. Parish aptly put it in his overview of the historiography of American slavery: "Much of the character of an institution may be revealed by its margins and its abnormalities. Exceptions may not prove rules but they can put them into clearer perspective." By illuminating slavery in the extremes, this book has attempted to explore the limits of slave family life and consider the best- and worst-case scenarios for parenting, family economies, marriage, and family stability. It has also clarified the nature and extent of agency in shaping slave family life—including its underlying similarities across the South. Indeed, at the risk of generalizing the results of this study it must be stated that while the extent of agency and its outer forms obviously varied across time and space, the *intent* of enslaved people with respect to their families appears to have been similar in each of the three regions studied here. Enslaved people in general attempted to maximize family time and contact (whether in the field, in the quarters, or in the form of weekend visiting), improve material conditions for themselves and their family members, and protect themselves and their family members from forced separation. The variations in all of these aspects in different regions and at different times had little to do with fundamental cultural variations but rather with the different boundaries and opportunities with which slaves in different parts of the South were confronted to realize their ideals.[2]

When analyzed side by side, the results of this interaction between the nature of regional agriculture and the behavior of slave families are most striking. The "typical"—for lack of a better term—slave family in Fairfax County was cross-plantation in structure, extremely vulnerable to forced separation, had an underdeveloped family economy, and encountered few opportunities for extensive family contact. These characteristics were clearly shaped by both external factors and slave agency. Cross-plantation marriages were not forced upon slaves from above but were rather slaves' solution to a structural demographic problem triggered by the decline in slave-based agriculture; after all, the slaves could have simply chosen not to marry or maintain long-term relationships. Forced separation plagued northern Virginia slave families, but many slaves did attempt to save themselves or their family members from separation, even if they were seldom successful. Likewise, the underdevelopment of slave family economies and lack of family contact were rooted in the nature of local agriculture, but again slaves clearly reacted to their situation—theft from the master and weekend visiting are good examples.

By contrast, the typical slave family in Georgetown District was co-residential in structure, less vulnerable to forced separation, had a highly developed family economy, and encountered more opportunities to maximize family contact. Unique external factors in the lowcountry stimulated unique reactions from enslaved people. Most chose to live in co-residential households where that was demographically possible, and the slave population as a whole made forced separation an unattractive financial proposition. Slaves also worked hard to complete their tasks early so as to maximize family contact and time to work for themselves in order to improve their material comforts. Less formidable boundaries and more opportunities led to greater long-term stability, more contact, and better material conditions than those of slaves living in northern Virginia.

Finally, the typical slave family living in St. James Parish was characterized by a co-residential domestic arrangement (but surrounded by a large number of singles), moderate vulnerability to forced separation, a moderately developed family economy, and very few opportunities for family contact. The work and social landscapes of enslaved people in southern Louisiana combined aspects found in both northern Virginia and lowcountry South Carolina, and were in many respects truly unique for North America. External factors weighed heavily indeed upon the slave population in the sugar country, but their actions certainly helped shape their family lives. As in Georgetown District, many chose to live in co-residential households where possible and expended tremendous energy in exploiting opportunities to acquire material goods for themselves and their families. Their lack of family contact cannot be attributed to a lack of cohesion but to gang labor and eighteen-hour shifts during the grinding season.

"Typical" slave families may have existed for northern Virginia, lowcountry South Carolina, and southern Louisiana, but a typical American slave family surely did not exist. The boundaries and opportunities with which families in different slave societies were confronted varied far too widely. In the aftermath of slavery, African-American families were confronted with new kinds of boundaries and opportunities, and their experiences during slavery no doubt influenced the way they seized new chances and dealt with new challenges. Families in different regions, however, drew from different family histories to forge new beginnings.

Notes

Abbreviations

APL   Alexandria Public Library, Alexandria, Virginia
FCRL   Fairfax City Regional Library
FWP   Federal Writers' Project
LC   Library of Congress, Washington, D.C.
LSU   Hill Memorial Library, Louisiana State University, Baton Rouge, Louisiana
LV   Library of Virginia, Richmond, Virginia
NARA National Archives and Records Administration, Washington, D.C.
RASP   Records of Ante-Bellum Southern Plantations: From the Revolution through the Civil War, edited by Kenneth M. Stampp (microfilm series).
SCHS   South Carolina Historical Society, Charleston, South Carolina
SPJC   St. James Parish Courthouse, Convent, Louisiana
USC   South Caroliniana Library, University of South Carolina, Columbia, South Carolina
VHS   Virginia Historical Society, Richmond, Virginia
WPA   Workers of the Writers' Program of the Works Progress Administration in the State of Virginia

Introduction

1. For a more detailed discussion of these shortcomings, see below.

2. Malone, *Sweet Chariot*, 7.

3. Phillips, *American Negro Slavery*, 342–43 and ch. 22; Fogel, *Without Consent or Contract*, 162–68.

4. Du Bois, *Negro American Family*, 18–25, 45–50, 99–104; Frazier, *Negro Family*, 23–30, 198–259; Stampp, *Peculiar Institution*, 343–48; Elkins, *Slavery*, 53–55; Moynihan, "Negro Family."

5. Blassingame, *Slave Community*, xi (quote); Fogel, *Without Consent or Contract*, 162–68.

6. The works of the revisionists are too numerous to list here. The most well known are Genovese, *Roll, Jordan, Roll*; Gutman, *Black Family*; Blassingame, *Slave Community*; Stuckey, *Slave Culture*.

7. John Blassingame was a notable exception. See Spindel, "Assessing Memory," 251.

8. Gutman, *Black Family*, 13, 102, 304–305.

9. Fogel and Engerman, *Time on the Cross*, 49–51.

10. Kolchin, *American Slavery*,137; Parish, *Slavery*, 76 (quote); Smith, *Debating Slavery*, 46–51; Dunaway, *African-American Family*, 4–5.

11. Fogel and Engerman's work is most aptly countered in David et al., *Reckoning with Slavery*, especially chapters 3 and 4. See also Gutman, *Slavery and the Numbers Game*, 88–164.

12. Berlin, *Generations of Captivity*, 4; Dunaway, *African-American Family*, 4–5.

13. Parish, *Slavery*, 97 (first quote); Morgan, *Slave Counterpoint*, xvii (second quote).

14. Hudson, *To Have and to Hold*; Malone, *Sweet Chariot*; West, *Chains of Love*; Berry, *Swing the Sickle*. Generalized conclusions of localized research, for example, can be found in the works of Malone, *Sweet Chariot*, 5; Stevenson, *Life*, 160–61; and Dunaway, *African-American Family*, 5.

15. Despite the differences in terminology, *county*, *district*, and *parish* all correspond in this case to what in most American states is called a "county," or, the political division of a state into smaller provinces, each with their own local governments. Georgetown District, South Carolina, was renamed Georgetown County after the Civil War.

Chapter 1. Three Slave Societies of the Non-Cotton South

1. Berlin, *Generations of Captivity*, 9 (quote).

2. Olmsted, *A Journey*, 169 (first quote), 169 (second and third quotes).

3. Sweig, *Slavery in Fairfax County*, 5; Netherton et al., *Fairfax County*, 22–36; Morgan, *Slave Counterpoint*, 33–34; Gutheim, *Potomac*, 70 (quote).

4. Sweig, "Importation of African Slaves," 507–24; Netherton et al., *Fairfax County*, 22–36.

5. WPA, *Negro in Virginia*, 65; Netherton et al., *Fairfax County*, 161–70.

6. Craven, *Soil Exhaustion*, 72–73 (quote); Gray, *History of Agriculture*, 2:589–92; Walsh, "Plantation Management," 400–401.

7. WPA, *Negro in Virginia*, 65; Sprouse, *Mount Air*, 19–20; Netherton et al., *Fairfax County*, 152–57; Craven, *Soil Exhaustion*, 77–80.

8. Netherton et al., *Fairfax County*, 152–57; Gray, *History of Agriculture*, 2:602–608;

Craven, *Soil Exhaustion*, 77-80; Klingaman, "Significance of Grain," 275; Kulikoff, *Tobacco and Slaves*, 157-58.

9. Gray, *History of Agriculture*, 2:606-607; Netherton et al., *Fairfax County*, 166, 168 (first quote); Von Briesen, *Letters of Elijah Fletcher*, 8 (second quote).

10. Davis, *Travels*, 390 (first quote); R. M. Scott, Sr., Diary, 8 Sept. 1814, FCRL (second quote); *Alexandria Gazette*, 8 Jan. 1822 (third quote); *Phenix Gazette* (Alexandria), 13 Nov. 1828 (fourth quote); ibid., 4 Aug. 1828 (fifth quote).

11. Gray, *History of Agriculture*, 2:818-19; *Alexandria Gazette*, 7 July 1855; ibid., 20 Jan. 1853 (quote).

12. Janney, *Memoirs*, 29-30 (first quote); Craven, *Soil Exhaustion*, 79-121; Gray, *History of Agriculture*, 2:811-20; Kulikoff, *Tobacco and Slaves*, 157-58; Peterson, "Alexandria Market," 104-14; Walsh, "Plantation Management," 404; Netherton et al., *Fairfax County*, 152-70.

13. Craven, *Soil Exhaustion*, 79-121; Walsh, "Plantation Management," 404; Peterson, "Alexandria Market," 104-14; Gutheim, *Potomac*, 179; Poland, *Frontier to Suburbia*, 115-31; Netherton et al., *Fairfax County*, 152-70; R. M. Scott, Sr., Diary, 4 Oct. 1820, FCRL (quote).

14. Virginian, *Yankees in Fairfax County*, 3-24; Abbott, "Yankee Farmers," 56-63, 59 (first quote); Netherton et al., *Fairfax County*, 152-70; *Alexandria Gazette*, 30 Sept. 1847 (second quote); ibid., 5 Feb. 1849.

15. Abbott, "Yankee Farmers," 56-63; Virginian, *Yankees in Fairfax County*, 4, 5 (first and second quotes); Lyell, *Second Visit*, 207 (third quote).

16. U.S. Census, 1810-1860; Bancroft, *Slave-Trading*, 145; Craven, *Soil Exhaustion*, 140-41; Netherton et al., *Fairfax County*, 270; Gamble, *Sully*, 59.

17. Netherton et al., *Fairfax County*, 152-70; R. M. Scott, Jr., Diary, 2 Oct. 1846, FCRL (first quote); *Alexandria Gazette*, 11 Jan. 1853 (second quote); Tower, *Slavery Unmasked*, 50 (third quote); U.S. Census, 1860 Agriculture; Robert, "Lee the Farmer," 431-32.

18. Alston, *Rice Planter and Sportsman*, 46 (first quote), 67-68 (second quote).

19. Wood, *Black Majority*, 56-57; Morgan, *Slave Counterpoint*, 33.

20. Wood, *Black Majority*, 57-59, 58 (quote); Weir, *Colonial South Carolina*, 145; Coclanis, "Rice Prices," 532.

21. Schwalm, *Hard Fight*, 8-9; Morgan, *Slave Counterpoint*, 59, 77-79, 95 (quote); Coclanis, "Rice Prices," 532; Berlin, *Generations of Captivity*, 67-68; Morgan, "Slave Sales," 907, 917.

22. Wood, *Black Majority*, 59-62; Morgan, *Slave Counterpoint*, 68, 182-83; Joyner, *Down by the Riverside*, 13-14, 57-59; Littlefield, *Rice and Slaves*, 74-114; Schwalm, *Hard Fight*, 9; Carney, *Black Rice*, ch. 3.

23. Wood, *Black Majority*, 87-91; Schwalm, *Hard Fight*, 13.

24. Rogers, *Georgetown County*, 27-29; Berlin, *Generations of Captivity*, 70-71; Joyner, *Down by the Riverside*, 41; Schwalm, *Hard Fight*, 8.

25. Morgan, *Slave Counterpoint*, 33–34; Gray, *History of Agriculture*, 2:593 (quote); Rogers, *Georgetown County*, 257; Weir, *Colonial South Carolina*, 335.

26. Kolchin, *American Slavery*, 72–73; Chaplin, "Tidal Rice," 38; Berlin, *Generations of Captivity*, 124–25; Rogers, *Georgetown County*, 342–43; Morgan, *Slave Counterpoint*, 62.

27. Gray, *History of Agriculture*, 2:721–22; Chaplin, "Tidal Rice," 31–34, 39; U.S. Census, 1800–1810; Schwalm, *Hard Fight*, 8 (quote); Swan, "Structure and Profitability," 322.

28. Rogers, *Georgetown County*, 324; Gragg, *Pirates, Planters, and Patriots*, 47–48 (quote); Dusinberre, *Them Dark Days*, 387–416.

29. Russell, *My Diary*, 1:192 (first quote); Pringle, *Chicora Wood*, 53 (second quote), 63 (third quote).

30. Michie, *Richmond Hill Plantation*, 40; Rogers, *Georgetown County*, 258, 259 (first quote), 261–62, 279, 324, 339; Ben Horry, in FWP, *Slave Narratives*, vol. 14, pt. ii, 309 (second quote).

31. Mills, *Statistics*, 558 (quote); Joyner, *Down by the Riverside*, 41.

32. Berlin, *Generations of Captivity*, 179; Rodrigue, *Reconstruction*, 9–10.

33. Sitterson, *Sugar Country*, 13–14; Whitten, *Andrew Durnford*, 19–20; Follett, *Sugar Masters*, 10–11; Le Gardeur, "Origins of the Sugar Industry," 4.

34. Sitterson, *Sugar Country*, 13 (first quote), 14; Follett, *Sugar Masters*, 10–11; Rodrigue, *Reconstruction*, 10; Le Gardeur, "Origins of the Sugar Industry," 4; Whitten, *Andrew Durnford*, 19–20; Aime, *Plantation Diary*, 101; Leon, *On Sugar Cultivation*, 10 (second quote).

35. Sitterson, *Sugar Country*, 14; Thorpe, "Sugar Region," 750 (first quote); Gray, *History of Agriculture*, 1:62–84; Moody, *Slavery*, 1, 7–8; Persac, "Plantations on the Mississippi River," LC; Tixier, *Tixier's Travels*, 41–42; Le Gardeur, "Origins of the Sugar Industry," 1–2 (second quote); Bourgeois, *Cabanocey*, 35.

36. Gray, *History of Agriculture*, 2:739; Le Gardeur, "Origins of the Sugar Industry," 4–9; Moody, *Slavery*, 9; Thorpe, "Sugar Region," 747–48; Sitterson, *Sugar Country*, 6–7.

37. Gray, *History of Agriculture*, 2:739–40; Sitterson, *Sugar Country*, 3; Le Gardeur, "Origins of the Sugar Industry," 9–22.

38. Sitterson, *Sugar Country*, 1–12; Moody, *Slavery*, 8–10; Rodrigue, *Reconstruction*, 10–11; Follett, *Sugar Masters*, 17–18; Gayarré, "Louisiana Sugar Plantation," 607 (quote).

39. Follet, *Sugar Masters*, 26–37; Sitterson, *Sugar Country*, 9–11; Leon, *On Sugar Cultivation*, 70 (first quote); Olmsted, *A Journey*, 661 (second quote).

40. Olmsted, *A Journey*, 661 (first quote); Sitterson, *Sugar Country*, 18–22; Leon, *On Sugar Cultivation*, 10 (second quote); Gray, *History of Agriculture*, 2:740; Moody, *Slavery*, 10–11; Tixier, *Tixier's Travels*, 54 (second quote); Rodrigue, *Reconstruction*, 12; Begnaud, "Louisiana Sugar Cane Industry," 29 (third quote).

41. In Louisiana, the term "creole" was used to describe people of European and

African descent who were born in Louisiana. Whites and blacks who were born elsewhere in North America were called simply "Americans."

42. Moody, *Slavery*, 13–18; Berlin, *Generations of Captivity*, 88–96; Din, *Spaniards, Planters, and Slaves*, 154–76, 186.

43. Berlin, *Generations of Captivity*, 179; Sitterson, *Sugar Country*, 10; Rodrigue, *Reconstruction*, 11; *Louisiana State Gazette*, 13 Feb. 1826 (second quote); *New Orleans Daily Picayune*, 12 Oct. 1851 (third quote).

44. Gray, *History of Agriculture*, 2:740–9; Follett, *Sugar Masters*, 20–21, 26–27; Sitterson, *Sugar Country*, 28–30; Begnaud, "Louisiana Sugar Cane Industry," 31; *De Bow's Review*, 8 (Jan. 1850): 35; Leon, *On Sugar Cultivation*, 3–5.

45. Johnston, *Letter*, 8; Gray, *History of Agriculture*, 2:1033; Follett, *Sugar Masters*, 18–23; W. C. C. Claiborne, in Rowland, *Official Letter Books*, 3:363 (first quote); Rodrigue, *Reconstruction*, 11–12; Sitterson, *Sugar Country*, 23–24; Pierre C. de Laussat, in Sitterson, *Sugar Country*, 23 (second quote).

46. Moody, *Slavery*, 5–6; Bourgeois, *Cabanocey*, 26–27, 38; U.S. Census, 1810; Schmitz, "Economies of Scale," 959–80; Menn, *Large Slaveholders*, 39, 112–13; U.S. Census, 1860, Slave Schedules; Pritchard, "Tourist's Description," 14; Thorpe, "Sugar Region," 759; "List of Slaves found on the Homestead of W. P. Welham," 10 Dec. 1860, Welham Plantation Record Books, LSU.

47. Aime, *Plantation Diary*, 8, 168; Toledano, "Louisiana's Golden Age," 217; Sitterson, *Sugar Country*, 19, 29; *De Bow's Review* 14 (Mar., 1853), 200 (quote); Champomier, *Statement of the Sugar Crop: 1849–60*.

48. Thorpe, "Sugar Region," 750 (first quote); Parker, *Trip to the West*, 227 (second quote); Tixier, *Tixier's Travels*, 53 (third quote); Tower, *Slavery Unmasked*, 287 (fourth quote); Pritchard, "Tourist's Description," 13 (fifth quote), 14.

Chapter 2. The Nature of Agricultural Labor

1. Perdue et al., *Weevils in the Wheat*, 25–26, 25 (quote).

2. The following description of tobacco cultivation is based on these sources: Morgan, *Slave Counterpoint*, 164–70; Gutheim, *Potomac*, 70–74; Breen, *Tobacco Culture*, 46–58.

3. Morgan, *Slave Counterpoint*, 187–94, 169 (first quote), 191 (second quote).

4. Unless otherwise mentioned, the description of the cultivation of wheat, corn, rye, and oats in this and the following paragraphs is based on David Wilson Scott, Diary, 1819–1821, Papers of David Wilson Scott, LC. See also Morgan, *Slave Counterpoint*, 170–78; Stevenson, *Life*, 191; Gill, "Wheat Culture," 380–93; Janney and Janney, *Janney's Virginia*, 68–76, 72–73 (quote).

5. Von Briesen, *Letters of Elijah Fletcher*, 21; Janney and Janney, *Janney's Virginia*, 72.

6. Janney and Janney, *Janney's Virginia*, 73–75, 74 (first quote); Drew, *North-Side*, 70 (second quote); Stevenson, *Life*, 191.

7. Janney and Janney, *Janney's Virginia*, 70–71, 71 (quote); Bushrod Washington to son, 27 July 1829, Papers of Bushrod Washington, LC.

8. D. W. Scott, Diary, LC; *Alexandria Gazette and Daily Advertiser*, 5 Jan. 1822 (first quote); Pryor, *Walney*, 52 (second quote).

9. Stevenson, *Life*, 197; Frobel, *Civil War Diary*, 82 (first quote), 113 (second quote), 123, 40–41 (third and fourth quotes).

10. Morgan, *Slave Counterpoint*, 190–91; Perdue et al., *Weevils in the Wheat*, 26; Stevenson, *Life*, 187–88; Berlin, *Generations of Captivity*, 212.

11. Craven, *Soil Exhaustion*, 114; D. W. Scott, Diary, LC; Morgan, *Slave Counterpoint*, 191 (first quote); Berlin, *Generations of Captivity*, 212.

12. Walsh, "Plantation Management," 405 (second quote), 406; WPA, *Negro in Virginia*, 66.

13. Stevenson, *Life*, 187, 189 (first quote); Steward, *Twenty-Two Years*, 4 (second quote); Drew, *North-Side*, 67–68 (third quote); George Jackson, in FWP, *Slave Narratives*, vol. 12, 46 (fourth quote); Von Briesen, *Letters of Elijah Fletcher*, 8, 14 (fifth quote).

14. Machen, *Letters*, 97 (first quote); Drew, *North-Side*, 156 (second quote).

15. Joyner, *Down by the Riverside*, 45–46; Pringle, *Chicora Wood*, 15 (first quote); Schwalm, *Hard Fight*, 14; Alston, *Rice Planter*, 44 (second quote).

16. Unless otherwise stated, the following description of rice cultivation is based on these sources: Gray, *History of Agriculture*, 2:726–31; Rogers, *Georgetown County*, 331–34; Schwaab, *Travels in the Old South*, 1:13–16; Phillips, *Plantation and Frontier*, 1:259–65; Doar, *Rice Planting*, 13–20; Schwalm, *Hard Fight*, 19–28; Joyner, *Down by the Riverside*, 45–50.

17. See Gray, *History of Agriculture*, 2:727; Schwaab, *Travels in the Old South*, 1:13; Phillips, *Plantation and Frontier*, 1:259 (quote).

18. Ben Horry, in FWP, *Slave Narratives*, vol. 14, part ii, 302.

19. Phillips, *Plantation and Frontier*, 1:262 (first quote); Hodgson, *Remarks*, 116 (second quote).

20. Schwalm, *Hard Fight*, 27; Pringle, *Chicora Wood*, 14 (first quote); Phillips, *Plantation and Frontier*, 1:263 (second quote); Rogers, *Georgetown County*, 335; Alston, *Rice Planter*, 42–43, 125; Maggie Black, in FWP, *Slave Narratives*, vol. 14, part i, 58–59 (third and fourth quotes); Joyner, *Down by the Riverside*, 48–49 (fifth quote).

21. Morgan, *Slave Counterpoint*, 179; Berlin, *Generations of Captivity*, 77–78; Hudson, *To Have and to Hold*, 2; Joyner, *Down by the Riverside*, 43; Rhyne, *Voices*, 107 (first quote); Collins, *Memories*, 107 (second quote); Nevins, *American Social History*, 154 (third quote); Olmsted, *A Journey*, 434 (fourth quote).

22. Morgan, "Work and Culture," 568–69 (first quote); Fogel, *Without Consent or Contract*, 193; Schwalm, *Hard Fight*, 13–14.

23. Berlin, *Generations of Captivity*, 77–78, 78 (first quote); Pringle, *Chicora Wood*, 67 (second quote); Alston, *Rice Planter*, 8 (third quote); Rhyne, *Voices*, 29 (fourth quote).

24. Berlin, *Generations of Captivity*, 77–78, 78 (quote); Carney, *Black Rice*, 99–100; Morgan, *Slave Counterpoint*, 181–83; Schwalm, *Hard Fight*, 13.

25. Hudson, *To Have and to Hold*, 4 (first quote); Olmsted, *A Journey*, 436 (second and third quotes). By "stampede to the 'swamp,'" Olmsted refers to an increase in the number of runaways. For "customary rights" of the enslaved, see Genovese, *Roll, Jordan, Roll*, 31–32; and in the lowcountry specifically, Schwalm, *Hard Fight*, 37–38.

26. Morgan, *Slave Counterpoint*, 183–85; Schwalm, *Hard Fight*, 14; Joyner, *Down by the Riverside*, 44; Rogers, *Georgetown County*, 331; Rhyne, *Voices*, 29 (first quote); Alston, *Rice Planter*, 46 (second quote); Olmsted, *A Journey*, 435 (third quote). See also Hudson, *To Have and to Hold*, 2–3; Morgan, "Work and Culture," 570; Gabe Lance, in FWP, *Slave Narratives*, vol. 14, part iii, 93. Rice planter J. Motte Alston did not assign task work during the harvest but rather worked his slaves from "early morn till late in the night" during that time. Alston, *Rice Planter*, 47.

27. Joyner, *Down by the Riverside*, 43; Weston, "Rules," in Collins, *Memories*, 107–108 (first quote); Easterby, *South Carolina Rice Plantation*, 346 (third quote).

28. Hudson, *To Have and to Hold*, 2; Schwaab, *Travels in the Old South*, 1:8 (first and second quotes).

29. Schwaab, *Travels in the Old South*, 1:8 (first quote); Hodgson, *Remarks*, 118 (second quote); Doar, *Rice Planting*, 33 (third quote); Nevins, *American Social History*, 154 (fourth quote); Olmsted, *A Journey*, 435 (fifth quote).

30. Alston, *Rice Planter*, 46 (quote).

31. Joyner, *Down by the Riverside*, 65–68; Phillips, *Plantation and Frontier*, 1:120 (first and second quotes).

32. Ben Horry, in FWP, *Slave Narratives*, vol. 14, part ii, 304 (first quote); Gabe Lance, in ibid., vol. 14, part iii, 92 (second quote); Albert Carolina, in ibid., vol. 14, part i, 197–98 (third quote); Morgan, "Work and Culture," 583 (fourth quote).

33. Tower, *Slavery Unmasked*, 286–87 (first quote); Albert, *House of Bondage*, 4 (second quote); Pritchard, "Routine," 168–69.

34. Aime, *Plantation Diary*, 135 (quote); Octave Colomb Plantation Journal, 23 Dec. 1850, RASP, Series H, LSU; H. M. Seale, Diary, 1853–1857, LSU; Benjamin Tureaud, Plantation Journal, 1854, Benjamin Tureaud Family Papers, LSU; Ashland Plantation Record Book, LSU; Pritchard, "Routine," 169; Olmsted, *A Journey*, 665; Northup, *Twelve Years a Slave*, 209; Gray, *History of Agriculture*, 2:749–50; Thorpe, "Sugar Region," 756; McDonald, "Independent Economic Production," 276–77.

35. Northup, *Twelve Years a Slave*, 209 (quote); Olmsted, *A Journey*, 665; Pritchard, "Routine," 169; McDonald, "Independent Economic Production," 276–77; Aime, *Plantation Diary*; Octave Colomb Plantation Journal, RASP, Series H, LSU; Seale, Diary, LSU; Benjamin Tureaud Plantation Journal, Benjamin Tureaud Family Papers, LSU.

36. Aime, *Plantation Diary*, 135–36; Octave Colomb Plantation Journal, RASP, Series H, LSU; Thorpe, "Sugar Region," 754–55 (first and second quotes); Robinson, *Solon Robinson*, 177; Pritchard, "Routine," 169–70.

37. Aime, *Plantation Diary*, 136–39, 136 (first quote); Octave Colomb Plantation Journal, RASP, Series H, LSU; Benjamin Tureaud Plantation Journal, Benjamin Tureaud Family Papers, LSU; H. M. Seale, Diary, 1853–1857, LSU; Thorpe, "Sugar Region," 756–57, 757 (second quote); Pritchard, "Routine," 170–71; Gray, *History of Agriculture*, 2:750; Olmsted, *A Journey*, 666–67.

38. Pritchard, "Routine," 171–73; McDonald, "Independent Economic Production," 277; Moody, *Slavery*, 46–47; Tixier, *Tixier's Travels*, 51; Aime, *Plantation Diary*, 97, 136–39; Octave Colomb Plantation Journal, RASP, Series H, LSU; H. M. Seale, Diary, 1853–1857, LSU; Benjamin Tureaud Plantation Journal, Benjamin Tureaud Family Papers, LSU; Thorpe, "Sugar Region," 758 (quote).

39. Pritchard, "Routine," 173–74; Thorpe, "Sugar Region," 759 (quote); Follett, *Sugar Masters*, 24; Olmsted, *A Journey*, 668; Schmitz, "Economies of Scale," 961.

40. Pritchard, "Routine," 173–75; Northup, *Twelve Years a Slave*, 209–11; Robinson, *Solon Robinson*, 159; Thorpe, "Sugar Region," 760–61 (quote); McDonald, "Independent Economic Production," 279; Aime, *Plantation Diary*, 139–141, 140; Octave Colomb Plantation Journal, 17 Nov. 1850, RASP, Series H, LSU; Seale, Diary, LSU; Benjamin Tureaud Plantation Journal, Benjamin Tureaud Family Papers, LSU; Ashland Plantation Record Book, 25 Dec. 1852, LSU.

41. Northup, *Twelve Years a Slave*, 211–13; Olmstead, *A Journey*, 670–73; Pritchard, "Routine," 175–77; Thorpe, "Sugar Region," 762–63; Rodrigue, *Reconstruction*, 14–15.

42. Follett, *Sugar Masters*, 93 (first quote); Berlin, *Generations of Captivity*, 180 (second quote).

43. Follett, *Sugar Masters*, 92–95; Berlin, *Generations of Captivity*, 180–81; McDonald, "Independent Economic Production," 278; Moody, *Slavery*, 45–53; Fogel, *Without Consent or Contract*, 25–26; Blassingame, *Slave Testimony*, 394–95.

44. Ingraham, *South-West*, 1:241; Sitterson, "Magnolia Plantation," 199–200; Northup, *Twelve Years a Slave*, 209–10; Olmsted, *A Journey*, 666–67; Follett, *Sugar Masters*, 92–97; Moody, *Slavery*, 45–48; Bauer, *Leader among Peers*, 51–52; Russell, *My Diary*, 1:379–80; Rodrigue, *Reconstruction*, 17–18; Thorpe, "Sugar Region," 757–58 (quote).

45. Parker, *Trip to the West*, 227 (first quote); Pritchard, "Tourist's Description," 12 (second quote); Robinson, *Solon Robinson*, 167 (third quote).

46. Clayton, *Mother Wit*, 129 (first quote), 65 (fourth quote), 84 (fifth quote); Drake, *Pictures of the "Peculiar Institution,"* 8 (second quote); Still, *Underground Railroad*, 403 (third quote); Wingfield, "Sugar Plantations," 46 (sixth quote).

47. Ripley, *Social Life*, 195 (first quote); Octave Colomb Plantation Journal, 5 Sept. 1852, RASP, Series H, LSU (second quote); Olmsted, *A Journey*, 651 f.n.; Clayton, *Mother Wit*, 84 (third quote), 107 (fourth quote); Aime, *Plantation Diary*, 97 (fifth quote).

48. Follett, *Sugar Masters*, 97 (quote), 106; Thorpe, "Sugar Region," 761; Rodrigue, *Reconstruction*, 15–16.

49. Octave Colomb Plantation Journal, 26 Nov. 1850, RASP, Series H, LSU (first quote); Moody, *Slavery*, 52; Russell, *My Diary*, 1:374–75 (second quote); Ingraham, *South-West*, 240 (third quote); Olmsted, *A Journey*, 668–69, 668 (fourth quote); Hamilton, *Men and Manners*, 2:229.

50. Hamilton, *Men and Manners*, 2:229 (first quote); Stirling, *Letters from the Slave States*, 126 (second quote); Drake, *Pictures of the "Peculiar Institution,"* 8 (third quote).

Chapter 3. Family Contact during Working Hours

1. Parts of this section were previously published, in adapted form, in Pargas, "Work and Slave Family Life." King, *Stolen Childhood*, 4.

2. Stevenson, *Life*, 104, 193 (first quote), 250–51.

3. Stevenson, *Life*, 104, 193, 250–51; McMillen, *Southern Women*, 72–73; Dunaway, *African-American Family*, 132; Drew, *North-Side*, 71 (second quote).

4. Sweig, "Northern Virginia Slavery," 107, 142.

5. Von Briesen, *Letters of Elijah Fletcher*, 8, 14; Blassingame, *Slave Community*, 179; Perdue et al., *Weevils in the Wheat*, 185 (quote).

6. Blassingame, *Slave Community*, 180–81; Drew, *North-Side*, 156 (first quote); Stevenson, *Life*, 250–51, 250 (second quote); Perdue et al., *Weevils in the Wheat*, 150 (third quote).

7. Drew, *North-Side*, 155 (first quote); Estate Inventory of M. C. Fitzhugh, Will Book Z-1, p. 73, FCRL; Perdue et al., *Weevils in the Wheat*, 25–26 (second quote).

8. Calculated from the U.S. Census, 1840. Figures for David Wilson Scott based on McMillon and Wall, *Fairfax County*, 18.

9. Sterling, *We Are Your Sisters*, 6; Drew, *North-Side*, 68 (first quote); Conway, *Testimonies*, 5 (second quote); Olmsted, *A Journey*, 17 (third quote); Veney, *Narrative*, 7–8 (fourth and fifth quotes).

10. Sterling, *We Are Your Sisters*, 41; Owens, *This Species of Property*, 41–42; Drew, *North-Side*, 105 (first quote); R. M. Scott, Sr., Diary, FCRL (second quote).

11. Stevenson, *Life*, 197 (first quote); Redpath, *Roving Editor*, 194 (second quote); Steward, *Twenty-Two Years*, 10 (third quote); Still, *Underground Railroad*, 399 (fourth quote).

12. Stevenson, *Life*, 187–8; George Jackson, in FWP, *Slave Narratives*, vol. 12, 45–46 (first quote); Yetman, *Voices from Slavery*, 176 (second quote); cited in Stevenson, *Life*, 188 (third quote).

13. Steward, *Twenty-Two Years*, 3 (first quote); Perdue et al., *Weevils in the Wheat*, 26 (second and fourth quotes); Drew, *North-Side*, 156 (third quote).

14. Van Deburg, *Slave Drivers*, 5; Perdue et al., *Weevils in the Wheat*, 26 (quote).

15. Stevenson, *Life*, 191–92; Morgan, *Slave Counterpoint*, 173–74; White, *Ar'n't I a Woman?*, 121.

16. Dusinberre, *Them Dark Days*, 411–16; Berlin, *Generations of Captivity*, 210–11;

Kolchin, *American Slavery*, 114; Schwalm, *Hard Fight*, 42–43; West, *Chains of Love*, 95–96; Hudson, *To Have and to Hold*, 3; Joyner, *Down by the Riverside*, 45; Collins, *Memories*, 109 (first quote); Easterby, *South Carolina Rice Plantation*, 346 (second quote).

17. Dusinberre, *Them Dark Days*, 411–16; William Lowndes, Plantation Book, 1802–1822, Papers of William Lowndes, LC.

18. Collins, *Memories*, 115 (quote). Out of 4,408 enslaved boys and girls under the age of ten, only 501 (11 percent) did *not* live on slaveholdings with at least one enslaved woman between the ages of fifty-five and one hundred. Calculations based on U.S. Census, 1840.

19. Joyner, *Down by the Riverside*, 63, 78; Schwalm, *Hard Fight*, 31–32; Rhyne, *Voices*, 107 (first quote); "List of Negroes on True Blue Plantation, Waccamaw," Estate of Francis M. Weston, 11 July 1864, UNC (I am indebted to Janet Wright for bringing this document to my attention); Nevins, *American Social History*, 154 (second quote); Henry Brown, in FWP, *Slave Narratives*, vol. 14, part i, 119 (third quote); Alston, *Rice Planter*, 46 (fourth quote); Pringle, *Chicora Wood*, 90–91 (fifth quote).

20. Boyle and Fitch, *Georgetown County Slave Narratives*, 12 (first quote), 20 (second quote); "List of Negroes on True Blue Plantation, Waccamaw," Estate of Francis M. Weston, 11 July 1864, UNC.

21. Schwaab, *Travels in the Old South*, 1:8 (first quote); Olmsted, *A Journey*, 424 (second quote); Henry Brown, in FWP, *Slave Narratives*, vol. 14, part i, 119 (third quote); Pringle, *Chicora Wood*, 90 (fourth quote); Collins, *Memories*, 108.

22. Schwalm, *Hard Fight*, 28–30; Olmsted, *A Journey*, 424, 433 (first and second quotes); "List of Negroes on True Blue Plantation, Waccamaw," Estate of Francis M. Weston, 11 July 1864, UNC.

23. Hudson, *To Have and to Hold*, 34–36; Schwalm, *Hard Fight*, 14 (first quote), 20–25; West, *Chains of Love*, 81 (second quote), 87–89; Easterby, *South Carolina Rice Plantation*, 346 (third quote), 270 (fourth quote); Olmsted, *A Journey*, 430 (fifth quote); Boyle and Fitch, *Georgetown County Slave Narratives*, 5–6 (sixth quote).

24. Schwaab, *Travels in the Old South*, 1:8 (first quote); Morgan, "Ownership of Property by Slaves," 402 (second quote); Easterby, *South Carolina Rice Plantation*, 346 (third quote); Joyner, *Down by the Riverside*, 53.

25. Clayton, *Mother Wit*, 48 (first quote); H. M. Seale Diary, 26 July 1853 LSU; Follett, *Sugar Masters*, 73–75, 73 (second quote).

26. Follett, *Sugar Masters*, 73–75; Russell, *My Diary*, 1:397 (first quote, emphasis mine); Moody, *Slavery*, 85–86, 85 (second quote).

27. Moody, *Slavery*, 46, 85–86; Follett, *Sugar Masters*, 73–75; Ripley, *Social Life*, 193 (first quote); H. M. Seale, Diary, 1853–1857, LSU; Clayton, *Mother Wit*, 85 (second quote); Robinson, *Solon Robinson*, 2:203 (third quote).

28. Russell, *My Diary*, 1:380–86 (first quote); Yakubik and Méndez, *Beyond the Great House*, 24; Pulszky, *White, Red, Black*, 105 (second quote); Olmsted, *A Journey*, 657–58 (third quote).

29. Albert, *House of Bondage*, 4 (first quote); Ripley, *Social Life*, 194 (second quote); Follett, *Sugar Masters*, 72–74, 74 (sixth quote). According to Follett, enslaved women in the Louisiana sugar parishes had shorter birth intervals—sixteen months on average—than women living in the cotton regions of the U.S. South, who averaged thirty-four months from birth to conception.

30. The number of slave children born in St. James Parish during the census year 1850 was 151, while the number of slaves who died that year was 194, a difference of 28 percent. Calculated from the Historical Census Browser on the website of Fisher Library, University of Virginia (http://fisher.lib.virginia.edu/collections/stats/histcensus/index.html). Tadman, "Demographic Cost of Sugar," 1534–75; Berlin, *Generations of Captivity*, 181. On Welham's plantation, not only were there few slave children, but the number of children that imported slave women had was often much lower than the number of children local Creole slave women had. Southern and Maria, for example, a couple forty-five and thirty-eight years old, respectively, and imported from other slaveholding states, had only two children recorded in an 1860 inventory: Virginia (age 2) and Southern (age 1). A Creole couple named Zoé (age 36) and Jean (age unknown), had six children, varying in age from two months to fourteen years old. Migrant slave women thus had fewer children than locally born women. "List of Slaves Found on the Homestead of W. P. Welham," 10 Dec. 1860, Welham Plantation Record Books, Keller Family Plantation Records, LSU; "Liste des Négres de L'Habitation 'Constantia,'" 1855, Uncle Sam Plantation Papers, LSU; Tower, *Slavery Unmasked*, 312 (first quote).

31. Valcour Aime Slave Records, 1821–1850, RASP, Series H, LSU (first and second quotes); Olmsted, *A Journey*, 658 (third quote).

32. Of the total number of slave children listed in the 1840 census—1,433—only 523 (or 36.5 percent) lived on holdings that contained at least one elderly woman aged fifty-five or older. However, when adjusted to exclude those holdings that contained fewer than five slave children—which were most likely not sugar plantations, but rather households with domestic servants—the percentage of children who lived on plantations where elderly slave women were resident jumps to 55 percent (or 475 out of 866). Calculations derived from U.S. Census, 1840.

33. Tadman, "Demographic Cost of Sugar," 1534–75; Berlin, *Generations of Captivity*, 181; U.S. Census, 1840, 1850, and 1860 Slave Schedules. In 1850 the parish counted 834 slaves aged 40–49. Ten years later, these slaves should have been listed in the age category 50–59. By 1860, however, the latter category listed only 613 slaves, a decrease of 27 percent. The number of slaves aged 50–59 in 1850 was also reduced significantly by 1860, from 448 to 222, a decrease of 50 percent. See also Valcour Aime Slave Records, 1821–1850, RASP, Series H, LSU; Drake, *Pictures of the 'Peculiar Institution,'* 6 (quote); "List of Slaves Found on the Homestead of W. P. Welham," 10 Dec. 1860, Welham Plantation Record Books, Keller Family Plantation Records, LSU.

34. Thorpe, "Sugar Region," 759 (first quote); Ripley, *Social Life*, 193 (second quote);

Russell, *My Diary*, 1:380 (third quote); Clayton, *Mother Wit*, 52 (fourth quote), 65 (fifth quote).

35. Clayton, *Mother Wit*, 165 (quote), 52–53.

36. "List of Slaves Found on the Homestead of W. P. Welham," 10 Dec. 1860, Welham Plantation Record Books, Keller Family Plantation Records, LSU; Albert, *House of Bondage*, 14 (first quote); Clayton, *Mother Wit*, 129 (second quote); Pulszky, *White, Red, Black*, 2:105 (third quote).

37. Abdy, *Journal and Residence*, 3:14 (first quote); Clayton, *Mother Wit*, 52 (second quote), 99 (third quote), 162 (fourth quote), 65 (fifth quote).

38. Yakubik and Méndez, *Beyond the Great House*, 24; Phillips, *American Negro Slavery*, 245; Clayton, *Mother Wit*, 99 (first quote); "List of Slaves Found on the Homestead of W. P. Welham," 10 Dec. 1860, Welham Plantation Record Books, Keller Family Plantation Records, LSU; Russell, *My Diary*, 1:396–97, 398 (second and third quotes).

39. Blassingame, *Slave Testimony*, 622–23; Yakublik and Méndez, *Beyond the Great House*, 16; Clayton, *Mother Wit*, 84 (first quote), 48 (third quote); Moody, *Slavery*, 46; Fred Brown, FWP, *Slave Narratives*, vol. 16, part i, 157 (second quote).

40. Ingraham, *South-West*, 242 (first quote); Hamilton, *Men and Manners*, 2:230 (second quote); Clayton, *Mother Wit*, 130 (third quote), 167 (fourth quote).

41. Moody, *Slavery*, 45–46; Rodrigue, *Reconstruction*, 18; Russell, *My Diary*, 1:379–80 (first and second quotes); Aime, *Plantation Diary*, 138 (third quote); Thorpe, "Sugar Region," 760–61; Octave Colomb Plantation Journal, 4 February 1850, RASP, Series H, LSU (fourth quote).

Chapter 4. Family-Based Internal Economies

1. Parts of this chapter were previously published, in different form, in Pargas, "'Various Means.'" Fogel, *Without Consent or Contract*, 189–94; Mintz, "The Question of Caribbean Peasantries," 31–34; Berlin, *Generations of Captivity*, 91.

2. Bolland, "Proto-Proletarians?," 140–41.

3. Nevins, *American Social History*, 392 (quotes).

4. Stevenson, *Life*, 186–87; Von Briesen, *Letters of Elijah Fletcher*, 14 (quote).

5. Stevenson, *Life*, 186–87; Steward, *Twenty-Two Years*, 3 (first quote); Yetman, *Voices from Slavery*, 176 (second quote).

6. Stevenson, *Life*, 186; George Jackson, FWP, *Slave Narratives*, vol. 12, 47 (first quote); Robinson, *Solon Robinson*, 237 (second quote).

7. Netherton et al., *Fairfax County*, 272, 212 (first quote); Fairfax County Deed Book, V-2, 123–24, APL; ibid., G-3, 147; Sweig, *Registrations of Free Negroes*.

8. Berlin, *Generations of Captivity*, 226 (first quote); George Jackson, FWP, *Slave Narratives*, vol. 12, 47 (second quote).

9. Stevenson, *Life*, 188–89; Netherton et al., *Fairfax County*, 21–22; Davis, *Travels*

of Four Years, 422 (second quote); Niemcewicz, *Under Their Vine and Fig Tree*, 89 (third quote).

10. Berlin, *Generations of Captivity*, 117, 226; Drew, *North-Side*, 155–56 (first and second quotes); Hudson, "'All that Cash,'" 84.

11. Olmsted, *A Journey*, 12 (first and fifth quotes); Sutcliffe, *Travels*, 37 (second, third, and fourth quotes); Hurst, *Alexandria on the Potomac*, 37.

12. George Jackson, FWP, *Slave Narratives*, vol. 12, 46 (first quote); Smith, *Autobiography*, 8 (second quote); Still, *Underground Railroad*, 391 (third quote); Niemcewicz, *Under the Vine and Fig Tree*, 100–101 (fourth quote); Stevenson, *Life*, 188; Steward, *Twenty-Two Years*, 3 (fifth quote).

13. Fogel, *Without Consent or Contract*, 192.

14. Olmsted, *A Journey*, 689 (quote); *Southern Cultivator*, cited in ibid.

15. Richard M. Scott's farm produced 500 bushels of wheat and 1,200 bushels of corn in 1850. Wheat sold in 1850 at approximately $1.01 per bushel, while corn sold at an average of $0.59 per bushel. Probably half of the corn crop, however, was consumed by Scott's family, slaves, and livestock. Records are incomplete, but I estimate that only 600 bushels were destined for the market. Scott had eight slaves between ten and fifty-five years old resident on the farm in 1850 (out of twenty resident slaves total). Thus, if 500 bushels of wheat and 600 bushels of corn were produced by eight hands, the gross return per able-bodied slave was approximately $107.38. Average annual expenses per able-bodied slave came to about $17, comprising the purchase of ready-made clothing ($14 per year according to a Fairfax County farmer who was interviewed by James Redpath in 1857), as well as bacon and herring (for which Scott paid $59.75 in 1850, or approximately $3 per slave). Expenses do not include cornmeal, which was produced on the farm. Thus, $107.38 minus $17 equals a net return of $90.38 per able-bodied slave. See U.S. Census, 1850 Agriculture and Slave Schedules; Gray, *History of Agriculture*, 2:1039 and vol. 1, 544; Taylor, "Feeding Slaves," 140–42; Redpath, *Roving Editor*, 193; R. M. Scott, Jr., Diary, 26 Aug. 1850, FCRL.

16. Machen, *Letters*, 32; Andrews, *Domestic Slave-Trade*, 169 (quote); Kolchin, *American Slavery*, 113–14; Stevenson, *Life*, 188–89.

17. Davis, *Travels of Four Years*, 423 (quote). For customary rights of the enslaved, see Genovese, *Roll, Jordan, Roll*, 30–31.

18. Hudson, "'All that Cash,'" 80–81; Fogel, *Without Consent or Contract*, 192–93; Netherton et al., *Fairfax County*, 157; Kolchin, *American Slavery*, 182; Von Briesen, *Letters of Elijah Fletcher*, 14 (quote).

19. Von Briesen, *Letters of Elijah Fletcher*, 14, 23, 26 (first and second quotes); Abdy, *Journal of a Residence*, 1:181 (third quote); Drew, *North-Side*, 155–57 (fourth and fifth quotes); General Assembly of Virginia, Petition of the Citizens of Fairfax County to the Delegates of the General Assembly of Virginia, 30 Dec. 1836, Legislative Petitions, LV (sixth and seventh quotes).

20. Olmsted, *A Journey*, 689 (first quote); Anonymous, "A Slave's Story," 617 (second and third quotes); Schlotterbeck, "Internal Economy," 171.

21. Alston, *Rice Planter*, 45 (quote); Rogers, *History of Georgetown County*, 348; Easterby, *South Carolina Rice Plantation*, 34; Joyner, *Down by the Riverside*, 86–88; Schwalm, *Hard Fight*, 32–33.

22. West, *Chains of Love*, 100; Owens, *This Species*, 202; Alston, *Rice Planter*, 46 (first quote); William Oliver, in FWP, *Slave Narratives*, vol. 14, part iii, 219 (second and third quotes); Ben Horry, in FWP, *Slave Narratives*, vol. 14, part ii, 303, 324 (fourth and fifth quotes); Olmsted, *A Journey*, 484–85 (sixth quote); Collins, *Memories*, 114 (seventh quote).

23. Schwalm, *Hard Fight*, 61 (quote). John D. Magill, owner of Richmond Hill on the Waccamaw River, did not allow his slaves to cultivate family gardens. Rice production per slave and per acre at Richmond Hill was the lowest of all the Waccamaw plantations, perhaps the result of deliberate resistance by the enslaved population. See Joyner, *Down by the Riverside*, 26–28; Ellen Godfrey, in FWP, *Slave Narratives*, vol. 14, part ii, 159.

24. Ward produced four million pounds of rice in 1850 with 1,092 slaves, of whom 673 were of working age (ages ten to fifty-five). The price of rice per pound in 1850 was approximately 4.3 cents per pound. Four million pounds of rice at 4.3 cents per pound equals $172,000, divided by 673 equals a gross return of $255.57 per hand, minus an average of $12 in annual expenses equals $243.57. U.S. Census, 1850 Slave Schedules; Rogers, *Georgetown County*, 259, 339; Gray, *History of Agriculture*, 1:544.

25. Hudson, *To Have and to Hold*, 25; Henry Brown, in FWP, *Slave Narratives*, vol. 14, part i, 119 (first quote); William Oliver, in FWP, *Slave Narratives*, vol. 14, part iii, 219 (second quote); Schwaab, *Travels in the Old South*, 1:8 (third quote); Hogdson, *Remarks*, 117 (fourth quote); Nevins, *American Social History*, 154 (fifth quote); Malet, *Errand to the South*, 82 (sixth quote); Olmsted, *A Journey*, 439, 422.

26. Boyle and Fitch, *Georgetown County Slave Narratives*, 7 (first quote); Schwaab, *Travels in the Old South*, 1:239 (second quote).

27. Russell, *My Diary*, 1:192 (first quote); Margaret Bryant, in FWP, *Slave Narratives*, vol. 14, part i, 147 (second quote); Hudson, *To Have and to Hold*, 32–33 (third quote); West, *Chains of Love*, 99–102; Schwalm, *Hard Fight*, 59–60; Wood, *Women's Work, Men's Work*, 41.

28. Morgan, "Ownership of Property," 404; Morgan, "Work and Culture," 584 (first quote); Schwalm, *Hard Fight*, 33 (second quote); William Lowndes, Plantation Book, 1802–1822, Papers of William Lowndes, LC (third quote); Ben Horry, in FWP, *Slave Narratives*, vol. 14, part ii, 309 (fourth quote).

29. Joyner, *Down by the Riverside*, 52 (first quote); Easterby, *South Carolina Rice Plantation*, 350–52; Russell, *My Diary*, 1:373 (second quote); Bremer, *Homes of the New World*, 1:297 (third quote); Easterby, *South Carolina Rice Plantation*, 349–50 (fourth quote); Olmsted, *A Journey*, 439, 443.

30. Henry, *Police Control*, 79–95; Margaret Bryant, in FWP, *Slave Narratives*, vol.

14, part i, 147; Alston, *Rice Planter*, 55 (first quote); Russell, *My Diary*, 1:196 (second quote); Schwalm, *Hard Fight*, 62; Hudson, *To Have and to Hold*, 18–19; Easterby, *South Carolina Rice Plantation*, 350 (third quote).

31. Schwalm, *Hard Fight*, 14; Hudson, *To Have and to Hold*, 10–12, 11 (quote).

32. Hudson, *To Have and to Hold*, 10–12, 11 (first and second quotes); Easterby, *South Carolina Rice Plantation*, 349 (third quote).

33. Cited in Schwalm, *Hard Fight*, 60 (first quote); Dusinberre, *Them Dark Days*, 317 (second quote); Collins, *Memories*, 105 (third quote); Easterby, *South Carolina Rice Plantation*, 346.

34. Easterby, *South Carolina Rice Plantation*, 350 (first quote); Olmsted, *A Journey*, 422, 428 (third and fourth quotes); Russell, *My Diary*, 1:196 (fifth quote).

35. "Names of Negroes on Plantation, Jan. 1850," Octave Colomb, Plantation Journal, in RASP, Series H, LSU; "List of Slaves Found on the Homestead of W. P. Welham, 10 Dec. 1860," Welham Plantation Record Books, Keller Family Plantation Records, LSU; "A List of Hands on Conway, 1854," H. M. Seale, Diary, LSU; Wingfield, "Sugar Plantations of William J. Minor," 47 (quote); "Account with the Negroes, 25 Dec. 1849," Bruce, Seddon & Wilkins Plantation Records, LSU; McDonald, "Independent Economic Production," 285.

36. Follett, *Sugar Masters*, 196, 200; Olmsted, *A Journey*, 651 f.n.; Tixer, *Tixier's Travels*, 47 (first quote); Ripley, *Social Life*, 195 (second quote); McDonald, "Independent Economic Production," 284–85.

37. Benjamin Tureaud Plantation Journal, 1858, Tureaud Family Papers, LSU; "Account with Negroes for Wood and Work, 1850," Octave Colomb Plantation Journal, 1849–1866, LSU; Cashbook and Daybook, Bruce, Seddon and Wilkins' Plantation Records, LSU; McDonald, "Independent Economic Production," 283–84.

38. Thorpe, "Sugar Region," 737 (quote); Aime, *Plantation Diary*, 97; Cashbook and Daybook, Bruce, Seddon and Wilkins' Plantation Records, LSU; Octave Colomb Plantation Journal, RASP, Series H, LSU; Tureaud Plantation Journal, 1858, Tureaud Family Papers, LSU; McDonald, "Independent Economic Production," 284–85.

39. Kingsford, *Impressions of the West*, 47 (first quote); Clayton, *Mother Wit*, 164 (second quote); Russell, *My Diary*, 1:371; McDonald, "Independent Economic Production," 286–87.

40. Thorpe, "Sugar Region," 759. Frederick Law Olmsted, drawing upon government statistics, reported that average expenses for food and clothing in the sugar country during the 1850s came to approximately $30 per hand. See Olmsted, *A Journey*, 686.

41. Ingraham, *South-West*, 1:236 (first quote); Pulszky, *White, Red, Black*, 2:104 (second quote); Thorpe, "Sugar Region," 753 (third quote); Olmsted, *A Journey*, 682 (fourth quote).

42. Blassingame, *Slave Testimony*, 393 (first quote); Russell, *My Diary*, 1:371 (second quote); Olmsted, *A Journey*, 682; Clayton, *Mother Wit*, 52 (third quote).

43. McDonald, "Independent Economic Production," 280; Moody, *Slavery*,

64–65, 64 (first quote); Aime, *Plantation Diary*, 83 (second quote); "Accounts with the Slaves, 1858," Ledger, Tureaud Family Papers, LSU; Sitterson, *Sugar Country*, 98; Clayton, *Mother Wit*, 101 (third quote).

44. McDonald, "Independent Economic Production," 281; Sitterson, *Sugar Country*, 98–99.

45. "Accounts with the Slaves, 1858," Ledger, Tureaud Family Papers, LSU; Cashbook and Daybook, Bruce, Seddon, and Wilkins Plantation Records, LSU; Octave Colomb, Plantation Journal, RASP, Series H, LSU; Moody, *Slavery*, 67.

46. McDonald, "Independent Economic Production," 281; Moody, *Slavery*, 66; Olmsted, *A Journey*, 674–75; Blassingame, *Slave Testimony*, 395; Kingsford, *Impressions of the West*, 48 (first quote); Tixier, *Tixier's Travels*, 46 (second quote).

47. Berlin, *Generations of Captivity*, 184 (first quote); Follett, *Sugar Masters*, 134; Din, *Spaniards, Planters, and Slaves*, 177–78 (second quote).

48. Olmsted, *A Journey*, 660 (first quote); Bauer, *Leader among Peers*, 58; Follett, *Sugar Masters*, 153 (second quote).

49. Octavia George, in FWP, *Slave Narratives*, vol. 13, 111 (quote).

Chapter 5. Slaveholding across Time and Space

1. Kolchin, *American Slavery*, 78.
2. U.S. Census, 1810–1860, NARA; Virginia Gunnell Scott, Diary, FCRL; Sweig, *Slavery in Fairfax County*, 71–73.
3. Netherton et al., *Fairfax County*, 161; Cashin, "Landscape and Memory," 491.
4. Fairfax County Legislative Petition, 27 Dec. 1847, VHS (first quote); Netherton et al., *Fairfax County*, 152–59; Andrews, *Slavery*, 117 (second quote).
5. Netherton et al., *Fairfax County*, 170 (first quote); Redpath, *Roving Editor*, 192, 194 (second and fifth quotes); *Alexandria Gazette*, 29 Apr. 1846 (third quote); Craven, *Soil Exhaustion*, 90–91, 103, 114, 126–27, 152, 158.
6. Gray, *History of Agriculture*, 2:813–14; Gutheim, *Potomac*, 182–83; R. M. Scott, Jr., Diary, 18 March 1846, 21 Nov. 1848, FCRL (first quote).
7. Sweig, *Slavery in Fairfax County*, 33–36; McMillon and Wall, *Fairfax County*.
8. Hoffmann, "Map of Fairfax County," LC; Virginian, *Yankees in Fairfax County*, 14, 11 (first and second quotes).
9. Calculated from McMillon and Wall, *Fairfax County*; U.S. Census, 1850, Slave Schedules, NARA.
10. Kolchin, *American Slavery*, 101.
11. U.S. Census, 1800–1860, NARA; Teel, *1810 Census*. Almost all of the slave population increase occurred between 1800 and 1830, however, perhaps largely as a result of purchases. Between 1830 and 1860, the slave population only barely increased, confirming high mortality rates and low fertility rates, and between 1830 and 1840 it actually decreased slightly, caused perhaps partially by a series of epidemics that

broke out between 1836 and 1840. See Ricards and Blackburn, "Demographic History of Slavery," 217.

12. Morgan, *Slave Counterpoint*, 35–37, 77.

13. Schwalm, *Hard Fight*, 12; Alston, *Rice Planter*, 42–45, 58 (quotes), 57, 108; Doar, *Rice Planting*, 18.

14. Phillips, *Plantation and Frontier*, 1:263 (first quote); Alston, *Rice Planter*, 67–68 (second, third and fourth quotes), 127 (fifth quote), 57 (sixth quote).

15. Dusinberre, *Them Dark Days*, 288 (first quote), 285–91, 289; Rogers, *Georgetown County*, 328–29; U.S. Census, 1830–1860, NARA; Lachicotte, *Georgetown Rice Plantations*, 122.

16. Rogers, *Georgetown County*, 259–60 (quote), 287; U.S. Census, 1820–1860, NARA.

17. Schwaab, *Travels in the Old South*, 1:7; Bremer, *Homes of the New World*, 1:288 (second quote); Rogers, *Georgetown County*, 255–56.

18. Joyner, *Down by the Riverside*, 18–33; U.S. Census, 1860, Slave Schedules, NARA; 1825 Map of All Saints Parish, reprinted in Devereux, *Rice Princes*, inside cover; Rogers, *Georgetown County*, 253–69, 273–303.

19. U.S. Census, 1820, 1860 Slave Schedules, NARA; Mills, "Atlas," LC; Joyner, *Down by the Riverside*, 11, 17 (map); Alston, *Rice Planter*, 120–21 (first quote); Alexander Glennie, in Bull, *All Saints Church*, 27 (second quote).

20. Mills, "Atlas," LC; Rogers, *Georgetown County*, 280, 282, 273, 290–92; Dusinberre, *Them Dark Days*, 312; Bremer, *Homes of the New World*, 285 (first quote); Pringle, *Chicora Wood*, 156–57 (second quote).

21. Calculations based on U.S. Census, 1820, 1850 Slave Schedules, NARA. Note that the 1850 Georgetown slave population as a whole—including children and the elderly—indicated sexual imbalance (a male/female ratio of 0.90). This was due to differential mortality rates, not only in Georgetown District but in the lowcountry as a whole. See Ricards and Blackburn, "Demographic History of Slavery," 219–21.

22. U.S. Census, 1810–1860, NARA; Smith and Smith, *Cane, Cotton and Crevasses*, 81–84; Bourgeois, *Cabanocey*, 28.

23. Schmitz, "Economies of Scale," 959–80; Sitterson, *Sugar Country*, 158; Leon, *On Sugar Cultivation*, 3–4, 67–70, 69–70 (first quote); Bourgeois, *Cabanocey*, 28; Olmsted, *A Journey*, 661 (second quote).

24. Sitterson, *Sugar Country*, 157–61, 196–97. Sitterson found that sugar planting required more capital than any other southern cash crop, which explains the high proportion of partnerships. By 1859 a quarter of all Louisiana sugar plantations were jointly owned and operated. See Rodrigue, *Reconstruction*, 20; Begnaud, "Louisiana Sugar Cane Industry," 34–37; *De Bow's Review*, vol. 8, no. 1 (Jan. 1850): 35; Ingraham, *South-West*, 1:242 (second quote); Stuart, *Three Years*, 249 (third quote); Thorpe, "Sugar Region," 758 (fourth quote); Leon, *On Sugar Cultivation*, 58 (fifth quote); Follett, *Sugar Masters*, 27–39.

25. Menn, *Large Slaveholders*, 24; US Census, 1860 Agriculture, NARA; Pritchard, "Tourist's Description," 14.

26. Sitterson, *Sugar Country*, 50 (first quote); Schmitz, "Economies of Scale," 960 (second quote); Rodrigue, *Reconstruction*, 24 (third quote).

27. U.S. Census, 1850 Agriculture & 1850 Slave Schedules, NARA; Berlin, *Generations of Captivity*, 190 (quote); Sitterson, *Sugar Country*, 50.

28. U.S. Census, 1850–1860, Agriculture & Slave Schedules, NARA; Andrew Crane, Bills of Sale and Receipts, 4 Mar. 1849, 21 Sept. 1849, 4 Oct. 1849, 11 Jan. 1850, 21 Jan. 1850, 15 Mar. 1850, 1 May 1850, 25 May 1850, 3 Jan. 1851, 16 Jun. 1859, Andrew E. Crane Family Papers, LSU; *La Messager*, 30 April 1852.

29. Sitterson, *Sugar Country*, 158, 180; *Louisiana Gazette*, 8 Aug. 1806 (first quote); Evans, *Pedestrious Tour*, 326 (second quote); Gore, *Memories*, 136 (third quote); Thorpe, "Sugar Region," 759.

30. Bourgeois, *Cabanocey*, 26–28; Toledano, "Louisiana's Golden Age," 211–21; Aime, *Plantation Diary*, 8, 20, 72; Valcour Aime Slave Records, 1821–1850, in RASP, Series H, LSU; U.S. Census, 1820–1860, NARA; Sternberg, *River Road*, 290–92.

31. U.S. Census, 1810–1860, NARA.

32. U.S. Census, 1840–1860 Slave Schedules, NARA. For the six slaves who ran away from Samuel Fagot's plantation in 1855, see *La Messager*, 2 Jun. 1855 and 30 Jun. 1855.

33. U.S. Census, 1810–1860, NARA; Tadman, "Demographic Cost of Sugar," 1534–75, 1537 (first quote); Hamilton, *Men and Manners*, 2:222 (second quote). On low birth rates in the sugar country, see also Follett, *Sugar Masters*, ch. 2. Ira Berlin calculated that the fertility rate in St. James Parish was 0.54, or 40 percent lower than the fertility rates in cotton-producing counties in Alabama. See Berlin, *Generations of Captivity*, 181.

34. In 1850 there were 151 slave births in St. James Parish, and 194 slave deaths, a difference of 28 percent. Fairfax County counted 74 slave births and 29 slave deaths that year, a difference of 61 percent. Georgetown District listed 318 births for 1850, and 288 deaths, a difference of only 9 percent. Calculations based on data from the Historical Census Browser on the website of Fisher Library, University of Virginia (http://fisher.lib.virginia.edu/collections/stats/histcensus/index.html).

35. U.S. Census, 1810–1860, NARA; Tadman, "Demographic Cost of Sugar," 1549; Malone, *Sweet Chariot*, 28–29. If in St. James Parish in 1850 there were 43 more deaths than births, and 1850 was a normal year, then we should have expected the slave population to decrease from 7,724 in 1850 to 7,294 in 1860. It did not—it increased to 8,128. This suggests that planters made up the difference by importing more slaves (as many as 834, or 10 percent of the 1860 slave population) during the 1850s. Calculations based on data from the Historical Census Browser.

36. Tixier, *Tixier's Travels*, 41–42 (first quote); Martineau, *Retrospect*, 2:9 (second quote); Brady, *George Washington's Beautiful Nelly*, 215 (third quote); Robinson, *Solon Robinson*, 2:174 (fourth quote).

37. Champomier, *Statement of the Sugar Crop* 1846; Persac, "Plantations on the Mississippi River," LC; *Le Louisianais*, 9 Dec. 1865; *Le Messager*, 22 Feb. 1851.

38. Tadman, "Demographic Cost of Sugar," 1538 (quote); Berlin, *Generations of Captivity*, 179–80; Follett, *Slave Masters*, 46–54, 49 (quote). Calculations derived from the U.S. Census, 1820, 1850 Slave Schedules, NARA.

Chapter 6. Marriage Strategies and Family Formation

1. Parts of this chapter were previously published, in adapted form, in Pargas, "Boundaries and Opportunities." Davis, "Review of the Conflicting Theories," 100–103; Phillips, *American Negro Slavery*; Stampp, *Peculiar Institution*; Frazier, *Negro Family*.

2. Davis, "Review of the Conflicting Theories," 102–103; Gutman, *Black Family*, xxii-xxiii (quote); Fogel and Engerman, *Time on the Cross*, 49–51. For relatively recent studies that point to the widespread existence of female-headed households in the antebellum South, see Stevenson, *Life*, ch. 7 and 8; Hudson, *To Have and to Hold*, ch. 4; Dunaway, *African-American Family*, 5 and ch. 2; West, *Chains of Love*, ch. 2.

3. Blassingame, *Slave Community*, 164; West, *Chains of Love*, 45–50.; Fogel, *Without Consent or Contract*, 178–79.

4. Von Briesen, *Letters of Elijah Fletcher*, 23 (quote).

5. This supports Brenda Stevenson's conclusion that broad marriages and female-headed households were "especially prevalent" in antebellum northern Virginia. Stevenson, *Life*, 209 (quote).

6. Estate Inventory of George Triplett, Will Book O-1, 329, FCRL; Estate Inventory of Charles Guy Broadwater, Will Book P-1, 428, FCRL. For divided households in the 1850s see, for example, Estate Inventory of William H. Wrenn, Will Book Y-1, 163, FCRL; Estate Inventory of Cassius Carter, Will Book Y-1, 232, FCRL; Estate Inventory of John Davis, Will Book Y-1, 302, FCRL. In the estate accounts of Fairfax County slaveholders there are several more examples of female-headed households, too numerous to mention here.

7. Estate Inventory of Charles Smith, Will Book P-1, 349–50, FCRL; Slave Inventory of F. L. Lee estate, Will Book P-1, 15, FCRL; Estate Inventory of Sabret Scott, Will Book P-1, 140, FCRL.

8. Stevenson, *Life*, 209; Estate Inventory of Charles Henderson, Will Book M-1, 261, FCRL; Estate Inventory of Mastrom Cockerill, Will Book P-1, 122, FCRL (second quote); Estate Inventory of James Burke, Will Book O-1, 200, FCRL.

9. James Redpath, *Roving Editor*, 184–85 (first quote); Von Briesen, *Letters of Elijah Fletcher*, 23–24, 24 (second quote), 23 (third quote).

10. Camp, *Closer to Freedom*, 6, 12–34.

11. David Wilson Scott, Diary, 20 May 1820, LC (first quote); Davis, *Travels*, 400 (second quote); Camp, *Closer to Freedom*, 13–20, 20 (third quote).

12. Camp, *Closer to Freedom*, 28–30; Stevenson, *Life*, 222; West, *Chains of Love*, 50.

13. Meaders, *Advertisements*, 115 (first quote), 119 (second quote); Still, *Underground Railroad*, 442–43 (third quote).

14. George Jackson, in FWP, *Slave Narratives*, vol. 12, 45 (first quote); Redpath, *Roving Editor*, 199 (second quote). For more examples of local broad couples, see Stills, *Underground Railroad*, 260, 385, 396, 468, and 469.

15. Frobel, *Civil War Diary*, 36 (first quote), 72, 110; R. M. Scott, Jr., Diary, 10 Jun. 1849, FCRL (second quote). The Scott journals for Bush Hill mention several other cases of cross-plantation marriages among the resident slaves. See, for example, the entries for 7 Oct. 1824, 7 Oct. 1825, 27 Dec. 1826, and 19 Dec. 1847, among others. Only one marriage between two resident slaves at Bush Hill is ever mentioned, on 16 July 1854.

16. Deed of Sale from Thomas A. P. Jones to George Dobson, Deed Book R-3, 126 (microfilm), APL; Scott, Sr., Diary, 7 Oct. 1824, FCRL (first quote); Scott, Jr., Diary, 26 Aug. 1846, FCRL (second quote), 1 Mar. 1848 (third quote).

17. Stevenson, *Life*, 230–31; Perdue et al., *Weevils* 25 (quote); U.S. Census, 1850 and 1860 Slave Schedules (microfilm), NARA. In 1850, 698 out of 3,177 local slaves were classified as mulatto. In 1860 950 out of 3,116 slaves were mulatto.

18. Blassingame, *Slave Community*, 164; Meaders, *Advertisements*, 32 (first quote), 158 (fourth quote); Niccolls, "'Old John,'" 78 (second quote); Steward, *Twenty-two Years*, 3 (third quote).

19. Estate Inventory of William Fitzhugh, Will Book J-1, 285–94, FCRL; Sweig, "Northern Virginia Slavery," 115–23; Meaders, *Advertisements*, 4 (quote).

20. Estate Inventory of William Fitzhugh, Will Book J-1, 285–94, FCRL; Sweig, "Northern Virginia Slavery," 115–23.; Netherton et al., *Fairfax County*, 161. Brenda Stevenson considers the Fitzhugh singles to be kinless. Stevenson, *Life*, 212.

21. For more on slave naming patterns, see Gutman, *Black Family*, 187–92.

22. Estate Inventory of B. R. Davis, Will Book O-1, 2, FCRL.

23. Drew, *North-Side*, 154 (first quote); Stills, *Underground Railroad*, 479 (second quote). Frank Bell claimed to have grown up on a plantation near Vienna that contained over 150 slaves, but no holdings in late-antebellum Fairfax County were anywhere near that size. Perdue et al., *Weevils*, 25–26, 26 (third quote).

24. US Census, 1830, NARA; Davison McDowell, Second Account Book, 1811–1837, Davison McDowell Papers, USC.

25. Easterby, *South Carolina Rice Plantation*, 331–32.

26. Plantation Book of William Lowndes, 1802–1822, Papers of William Lowndes, LC; Slave Inventory of Paul D. Weston Estate, 1837, Paul D. Weston Papers, in RASP, Series B, LSU.

27. List of Weehaw People, 31 Dec. 1855, Henry A. Middleton, Jr., Weehaw Plantation Journal, 1855–1861, Cheves-Middleton Papers, SCHS; List of Negroes at the White House, May 1858, Julius Izard Pringle, White House Plantation Book, 1857, Pringle Family Papers, SCHS. See also Hudson, *To Have and to Hold*, 149.

28. List of Negroes belonging to Friendfield Estate, July 1841, James Ritchie

Sparkman Papers, 1822–1865, in RASP, Series A, pt. 2, LSU; Dirleton List of Negroes, 12 Feb. 1859, John Sparkman Plantation Book, 1859–1864, James Ritchie Sparkman Papers; List of Negroes from Northampton, Mar. 1853, James Ritchie Sparkman Papers; List of Negroes on True Blue Plantation, Emily F. Weston in Equity, 1864, UNC; List of Working People, Jan. 1860, Paul D. Weston Papers, RASP, Series B, LSU. Of the 242 slaves living on Friendfield plantation in 1841, 196 appear to have been living in households headed by co-residential couples (81 percent). Dirleton plantation counted twenty-one households headed by co-residential couples in 1859, accounting for ninety-seven of 166 slaves (58 percent). A total of twenty co-residential couples lived with their children at Northampton plantation in 1853; such households contained 103 of 120 slaves (86 percent). Likewise, a large majority of the slaves who resided on True Blue plantation on the Waccamaw in 1864, as well as those who belonged to the estate of Paul D. Weston in 1860, also lived in two-parent households (73 percent and 80 percent, respectively).

29. Bull, *All Saints' Church,* 32 (first quote); *Winyah Intelligencer,* 13 Mar. 1819, 6 Feb. 1830 (second quote), 30 Jul. 1830.

30. Sabe Rutledge, in FWP, *Slave Narratives,* vol. 14, part iv, 65; Gabe Lance, in FWP, *Slave Narratives,* vol. 14, part iii, 92 (first quote); Ben Horry, in FWP, *Slave Narratives,* vol. 14, part ii, 311 (second quote); William Oliver, in FWP, *Slave Narratives,* vol. 14, part iii, 217 (third quote); Rhyne, *Voices,* 13 (fourth quote).

31. Davison McDowell, Second Account Book, 1811–1837, Davison McDowell Papers, USC (first quote); Charles C. Pinckney Plantation Book, 1812–1861, Papers of Charles C. Pinckney, LC (second quote); Easterby, *South Carolina Rice Plantation,* 331–32; Plantation Book of William Lowndes, 1802–1822, Papers of William Lowndes, LC; Slave Inventory of Paul D. Weston Estate, 1837, Paul D. Weston Papers, in RASP, Series B, LSU.

32. List of Weehaw People, 31 Dec. 1855, Henry A. Middleton, Jr., Weehaw Plantation Journal, 1855–1861, Cheves-Middleton Papers, SCHS; List of Negroes belonging to Friendfield Estate, July 1841, James Ritchie Sparkman Papers, 1822–1865, in RASP, Series A, pt. 2, LSU; List of Negroes from Northampton, Mar. 1853, James Ritchie Sparkman Papers; Ben Horry, in FWP, *Slave Narratives,* vol. 14, part ii, 304 (quote).

33. Dirleton List of Negroes, 12 Feb. 1859, John Sparkman Plantation Book, 1859–1864, James Ritchie Sparkman Papers; Charles C. Pinckney Plantation Book, 1812–1861, Papers of Charles C. Pinckney, LC (quote).

34. In 1850 only 99 out of 17,894 slaves in Georgetown District were listed as mulatto. In 1860 some 236 out of 17,920 slaves were classified as mulatto. US Census, 1850 and 1860 Slave Schedules, NARA; Ben Horry, in FWP, *Slave Narratives,* vol. 14, part ii, 305 (quote). In Horry's quote, "thief children" refers to unwanted children, literally children stolen from the womb. "Woman overpower" means rape.

35. Doar, *Rice Planting,* 37 (first quote); Phillips, *Plantation and Frontier,* 1:116 (second quote); Henry, *Police Control,* 29; Joyner, *Down by the Riverside,* 132; Sam Polite, in FWP, *Slave Narratives,* vol. 14, part iii, 271 (third quote).

36. Rhyne, *Voices*, 108 (first quote); Olmsted, *A Journey*, 448 (second quote).

37. Doar, *Rice Planting*, 37 (first quote, italics mine); Davison McDowell, Second Account Book, 1811–1837, Davison McDowell Papers, USC; Mills, "Atlas," LC; Phillips, *Plantation and Frontier*, 1:116 (second quote, italics mine).

38. *Winyah Intelligencer*, 6 Mar. 1819 (first quote), 2 May 1829 (second quote), 7 Jan. 1829 (third quote).

39. Gore, *Memories*, 137 (quote).

40. Sitterson, *Sugar Country*, 90 (first quote); Pulszky, *White, Red, Black*, 2:107 (second quote).

41. Follett, *Sugar Masters*, 79; Olmsted, *A Journey*, 680 (first quote); Rodrigue, *Reconstruction*, 30–31 (second quote).

42. *La Messager*, 2 Jun. 1855, 16 Jun. 1855, 30 Jun. 1855; Still, *Underground Railroad*, 403–405, 403 (quote).

43. US Census, 1810, NARA; Succession of Michel Cantrelle, 1811–1816, Probate Records, SJPC.

44. U.S. Census, 1830, NARA; Succession of Daniel Blouin, 23 Nov. 1831, Probate Records, SJPC; Estate Inventory of Joseph Melancon, Jr., 10 Jan. 1837, Probate Records, SJPC.

45. *Planters' Advocate*, 23 May 1840; Succession of George Mather, 1 June 1837, Probate Records, SJPC.

46. The plantation and slaves of P. M. Lapice were seized by the sheriff and offered for sale following the court case *John Brown vs. P. M. Lapice*. The entire estate was advertised in *Le Messager* on 28 May 1852. Although the slaves are not advertised in family groups, they are listed in family order, for example: "Solomon (41), Charlotte (38), Robert (21), Kitty (14), Columbia (7), Caroline (4)" and so forth. I counted at least 156 slaves living in two-parent households. Where there was any doubt, I did not count them as two-parent households. "List of Slaves," 10 Dec. 1860, Welham Plantation Record Books, Keller Family Plantation Records, LSU; "Liste des Négres de L'Habitation 'Constancia,'" 1855, Uncle Sam Plantation Papers, LSU; "Register of Colored Persons Employed by or Attached to the Premises of Widow S. Fagot on Uncle Sam Plantation in the Parish of St. James in the State of Louisiana," 1865, Uncle Sam Plantation Papers, LSU; US Census, 1860 Slave Schedules, NARA; "Register of Colored Persons Employed by or Attached to the Premises of Madam Winchester on Buena Vista Plantation in the Parish of St. James in the State of Louisiana," 1865, Registers of Black Persons, Records of the Field Offices for the State of Louisiana, Bureau of Refugees, Freedmen, and Abandoned Lands, RG 105, NARA; "Register of Colored Persons Employed by or Attached to the Premises of James M. Morson on Wilton Plantation in the Parish of St. James in the State of Louisiana," 1865, ibid; "Register of Colored Persons Employed by or Attached to the Premises of P. J. B. Webre, Parish of St. James in the State of Louisiana," 1865, ibid.

47. Malone, *Sweet Chariot*, 54–55; H. M. Seale Diary, 20 Apr. 1853, 27 May 1853,

LSU (first quote); Pulszky, *White, Red, Black*, 2:106 (second quote); Russell, *My Diary*, 1:387 (third quote), 186 (fourth quote).

48. Fred Brown, in FWP, *Slave Narratives*, vol. 16, part i, 156 (first quote); Clayton, *Mother Wit*, 45 (second quote), 197 (third quote), 98 (third quote), 108 (fourth quote), 130 (fifth quote).

49. Malone, *Sweet Chariot*, 54–55.

50. See, for example, the slave lists in the following records: "Obligacion hipotecaria," 22 Jan. 1816, Succession of Michel Cantrelle, Probate Records, SJPC; Succession of Felicité Bourgeois, 10 Dec. 1823, Probate Records, SPJC; Estate Inventory of Daniel Blouin, Probate Records, SJPC.

51. Succession of George Mather, Probate Records, SPJC; *Planters' Advocate*, 23 May 1840; Follett, *Sugar Masters*, 68–69.

52. *Le Messager*, 28 May 1852; "Liste des Négres," Uncle Sam Plantation Papers, LSU; "Register of Colored Persons," ibid.; "List of Slaves," Welham Plantation Record Books, Keller Family Plantation Records, LSU.

53. George W. Jones to Jean Baptiste Landry, 9 Jun. 1850, Severin Landry Papers, LSU (first quote); Tixier, *Tixier's Travels*, 47 (second quote); Clayton, *Mother Wit*, 70 (third quote); Blassingame, *Slave Testimony*, 623 (fourth quote); Northup, *Twelve Years*, 221 (fifth quote).

54. Sitterson, *Sugar Country*, 103; Moody, *Slavery*, 95; Wingfield, "Sugar Plantations," 46 (first quote); George W. Jones to Jean Baptiste Landry, 9 Jun. 1850, Severin Landry Papers, LSU (second quote); Clayton, *Mother Wit*, 162 (third quote); Northup, *Twelve Years*, 221 (fourth quote).

55. Tixier, *Tixier's Travels*, 47 (first quote, italics mine); "List of Slaves," Welham Plantation Record Books, Keller Family Plantation Records, LSU; Persac, "Plantations on the Mississippi," LC; Sitterson, *Sugar Country*, 103 (second quote); Sitterson, "The McCollams," 347–67 (third quote).

56. Russell, *My Diary*, 1:397 (quote).

Chapter 7. Forced Separation

1. Portions of this chapter were previously published in adapted form in Pargas, "Disposing of Human Property." Fogel and Engerman, *Time on the Cross*, 49; Tadman, *Speculators and Slaves*, 134, 170–71, 178. For recent analyses of forced separation through local sales, long-term hiring, and estate divisions, see West, *Chains of Love*, 141–56; and Dunaway, *African-American Family*, 53–62.

2. Testament of Joshua Buckley, Will Book M-1, 249–50, FCRL; Testament of Peter Coulter, Will Book P-1, 242–43, FCRL. The will books are full of similar examples, far too numerous to mention here.

3. Testament of Levy Stone, Will Book N-1, 283, FCRL; Testament of Robert Gunnell, Will Book P-1, 173, FCRL; Testament of Ann Boggess, Will Book N-1, 380–81,

FCRL; Testament and Estate Inventory of William Watters, Will Book O-1, 399, Will Book P-1, 145, FCRL; Testament of P. J. Reid, Will Book W-1, 202, FCRL.

4. Netherton et al., *Fairfax County*, 217; Redpath, *Roving Editor*, 202 (quote). In 1852 one freed slave woman named Harriet Hambleton succeeded in buying her son Alfred for the sum of $800. See Deed Book R-3, 127, APL. Other slaves were bought by their free spouses, but it is unclear whether these free blacks had been born free or were manumitted.

5. Emancipation by John Compton, Deed Book X-2, 392, APL; Bequeathal by Robert Reid, Deed Book D-3, 182, APL.

6. *Alexandria Gazette*, 24 Aug. 1855 (first quote), 9 May 1822 (second quote); R. M. Scott, Sr., Diary, 27 Oct. 1827, FCRL.

7. Estate Account of Edward Blackburn, Will Book P-1, 286, FCRL; Estate Account of Wormley Carter, Will Book N-1, 261, FCRL; *Alexandria Gazette*, 29 Dec. 1851 (second quote).

8. Deyle, *Carry Me Back*, 41–46, 283–89; Gray, *History of Agriculture*, 2:65; Tadman, *Speculators and Slaves*, 170–72; Netherton et al., *Fairfax County*, 158–59; Sweig, "Northern Virginia Slavery," 191–93, 192–93; Bancroft, *Slave Trading*, 49.

9. In 1850, when the slave population was at 3,178, there were 74 births and 29 deaths. If 1850 was a typical year, the slave population should have grown to 3,628 by 1860. Instead it declined absolutely by 2 percent, from 3,178 to 3,116. Considering projected population growth, however, this amounted to a decline of 16 percent. Calculations based on data from the Historical Census Browser: http://fisher.lib.virginia.edu/collections/stats/histcensus/index.html.

10. Netherton et al., *Fairfax County*, 156; R. M. Scott, Sr., Diary, 4 Oct. 1820, FCRL (first quote); R. M. Scott, Jr., Diary, 2 Oct. 1846, FCRL (second quote); Drew, *North-Side*, 343 (third quote).

11. Netherton et al., *Fairfax County*, 263; *Phenix Gazette*, 25 Dec. 1828 (first quote); *Alexandria Gazette*, 24 Nov. 1859 (second and third quotes).

12. Estate Account of William Lane, Sr., Will Book P-1, 247, FCRL; Gamble, *Sully*, 65 (second quote); Estate Account of Ann Mason, Will Book Z-1, 153, FCRL; Estate Account of John Huntington, Will Book W-1, 145, FCRL; Estate Account of James Potter, Will Book X-1, 423, FCRL; Estate Account of Sarah McInteer, Will Book X-1, 120, FCRL.

13. Abdy, *Journal of a Residence*, 2:98–99, 177–78 (first and fourth quotes); Bancroft, *Slave-Trading*, 15 (second quote); Lawrence Lewis to Major Edward G. W. Butler, 18 Jan. 1837, Custis-Lee Family Papers, LC (fourth quote). R. M. Scott, Jr., Diary, 10–11 Aug. 1846 (sixth quote), 5 Oct. 1850, FCRL (seventh quote). See also entries on 17 Nov. 1852, 7 Sept. 1853, 17 Nov. 1853, 6 Feb. 1854, 17 Mar. 1854.

14. Conway, *Testimonies*, 21 (first quote); Still, *Underground Railroad*, 479 (second quote); Drew, *North-Side*, 29 (third quote).

15. Bancroft, *Slave-Trading*, 59; Andrews, *Slavery and the Domestic Slave Trade*, 137–39 (first, second and third quotes); Redpath, *Roving Editor*, 199 (fourth quote).

16. Estate Account of Elizabeth Tyler, Will Book O-1, 424, FCRL; *Phenix Gazette*, 31 Jan. 1826 (first quote); Franklin and Schweninger, *Runaway Slaves*, 391 (second quote), 396 (third quote); Blassingame, *Slave Testimony*, 87 (fourth and fifth quotes). See also Deyle, *Carry Me Back*, 245–75.

17. Drew, *North-Side*, 338 (quote).

18. Pryor, "Flexibility and Profit," 4–6; Bancroft, *Slave-Trading*, 145–46; Sweig, "Northern Virginia Slavery," 161 (quote).

19. Sweig, "Northern Virginia Slavery," 158–61, 158 (first quote); *Alexandria Gazette*, 25 May 1857 (second quote); Pryor, "Flexibility and Profit," 7; Bancroft, *Slave-Trading*, 145; Martin, *Divided Mastery*, 74–86, 75 (third quote).

20. Bancroft, *Slave-Trading*, 148; Sweig, "Northern Virginia Slavery," 159–60; Estate Account of George and Sarah Sweeney, Will Book X-1, 63, FCRL; Estate Account of Matilda B. West, Will Book N-1, 157, FCRL; R.M. Scott, Jr., Diary, FCRL.

21. R. M. Scott, Jr., Diary, 1 Jan. 1858, FCRL (first quote); *Alexandria Gazette*, 30 Dec. 1851 (second quote); Redpath, *Roving Editor*, 184–85, 188; Abdy, *Journal of a Residence*, 2:59; Andrews, *Slavery and the Domestic Slave-Trade*, 117–18; Allen, *Life and Letters*, 149; Olmsted, *A Journey*, 4–5.

22. Estate Account of John Fitzhugh, Will Book M-1, 390–98, FCRL; Estate Account of William Lane, Will Book O-1, 219, FCRL; Estate Account of Alfred Offret, Will Book N-1, 395, FCRL; Estate Account of William R. Newman, Will Book Z-1, 188, FCRL.

23. Sweig, "Northern Virginia Slavery," 161; Pryor, "Flexibility and Profit," 11; Robert, "Lee the Farmer," 435 (quote). One of Lee's slaves was at one point hired out in Alabama. See Blassingame, *Slave Testimony*, 467–68.

24. Martin, *Divided Mastery*, 138–60; Redpath, *Roving Editor*, 188; Sweig, "Northern Virginia Slavery," 161; Frobel, *Civil War Diary*, 215; R. M. Scott, Jr., Diary, 1 Mar. 1848, 1 Jan. 1849, FCRL (quote).

25. V. G. Scott, Diary, 25 Apr. 1857 (first quote), 4 May 1857, FCRL; Dziobek et al., "Mount Erin," 39 (second quote), 54 (third quote).

26. V. G. Scott, Diary, 14 July 1859 (first quote), 18 July 1859 (second quote), FCRL; Still, *Underground Railroad*, 391 (third quote); *Alexandria Gazette*, 7 Feb. 1853 (fourth quote).

27. Rogers, *Georgetown County*, 426; Ricards and Blackburn, "Demographic History," 215; Allston, *Allstons and Alstons*, 88 (first quote); Alston, *Rice Planter*, 56 (second quote); Oliphant, *Patriots and Filibusters*, 140 (third quote); Schwaab, *Travels in the Old South*, 1:10 (fourth quote).

28. Rogers, *Georgetown County*, 258–61, 267–68, see also ch. 13; Will of William Alston, 1838, Charleston County Will Book 41, 939, Charleston, S.C.; Trinkley, *Archaeological Study*, 52; Alston, *Rice Planter*, 47 (quote).

29. Easterby, *South Carolina Rice Plantation*, 28 (quote), 331–32; "List of Negroes belonging to my [Francis Withers'] Friendfield Estate, including Mount Pleasant, Midway and Canaan Plantations, July 1841," James Ritchie Sparkman Papers, RASP,

Series A, pt. 2, LSU; "List of Negroes for Northampton Estate, including Westfield and Bonny Neck, July 1841," ibid.; "List of Negroes Removed from Northampton and Appointed to this Place in March 1853," ibid.; James Ritchie Sparkman, Memorandum Book, 1846, ibid.; Gourdin, *Gourdin*, 16.

30. Dusinberre, *Them Dark Days*, 402–403.

31. *Winyah Intelligencer*, 20 Feb. 1830 (first quote), 3 Jan. 1829 (second quote), 6 Oct. 1832 (third quote).

32. Stephen D. Doar, Account Book 1, Papers of Stephen D. Doar, LC; Dusinberre, *Them Dark Days*, 402–403, 524; Easterby, *South Carolina Rice Plantation*, 337, 351–54; Slave Sale Broadside, 27 Jan. 1857, Cleland Kinloch Huger Papers, South Caroliniana Library, USC.

33. *Winyah Intelligencer*, 7 Jan. 1829 (first quote), 6 Feb. 1830 (second quote), 29 Sept. 1830 (third quote).

34. Hudson, *To Have and to Hold*, 175; Slave Inventory of the Paul D. Weston Estate, Paul D. Weston Papers, RASP, Series B, LSU; "Sales of Negroes of the Estate of E. J. Heriot," 4 Oct. 1859, James Ritchie Sparkman Papers, RASP, Series A, pt. 2 LSU; *Winyah Intelligencer*, 31 Mar. 1819 (first quote), 10 Nov. 1819 (third quote), 24 Nov. 1819 (fourth quote); Slave Sale Broadside, 27 Jan. 1857, and Slave Sale Broadside, 1 March 1854, James Ritchie Sparkman Papers (second quote).

35. Easterby, *South Carolina Rice Plantation*, 30 (first quote); Schwalm, *Hard Fight*, 56 (second, third and fourth quotes).

36. Pringle, *Chicora Wood*, 10–11 (quote).

37. Hudson, *To Have and to Hold*, 175; Rogers, *Georgetown County*, 343. The Georgetown slave population increased by 6,218 between 1800 and 1860, from 11,816 to 18,034. In 1850, according to vital statistics, the Georgetown slave population counted 318 births and 288 deaths—a difference of only 30. If 1850 was a normal year, the slave population should have grown by 300 by 1860; instead it grew by only 140 (from 17,894 in 1850 to 18,034 in 1860), with approximately 160 slaves unaccounted for. For vital statistics in 1850, see the Historical Census Browser at http://fisher.lib.virginia.edu/collections/stats/histcensus/index.html.

38. Bancroft, *Slave Trading*, 23; Deyle, *Carry Me Back*, 296; *Winyaw Intelligencer*, 9 Jan. 1830 (first quote); Welcome Beese, in FWP, *Slave Narratives*, vol. 14, part i, 49 (second quote); Lizzie Davis, in ibid., vol. 14, part i, 293 (third quote); Dusinberre, *Them Dark Days*, 403 (fourth quote).

39. *Winyaw Intelligencer*, 1 Dec. 1830 (quote); Tadman, *Speculators and Slaves*, 263.

40. Dusinberre, *Them Dark Days*, 298–99, 305, 402–403 (first, second, and third quotes); Easterby, *South Carolina Rice Plantation*, 29–46–48.

41. Easterby, *South Carolina Rice Plantation*, 337; *Winyaw Intelligencer*, 3 Jan. 1829 (quote).

42. Dusinberre, *Them Dark Days*, 300; Memorandum Book, James Ritchie Sparkman Papers, RASP, Series A, pt. 2, LSU; William Lowndes, Plantation Book, Pa-

pers of William Lowndes, LC (first quote); *Winyaw Intelligencer*, 1 Jan. 1825 (second quote).

43. *Winyaw Intelligencer*, 8 Dec. 1819 (first quote), 15 Dec. 1830 (second quote), 3 Jan. 1829 (third and fourth quotes).

44. See Sitterson, *Sugar Country*, 183 (quote).

45. Malone, *Sweet Chariot*, 213-16.

46. Succession of Leonard Fabre, 24 Mar. 1828, Probate Records, SJPC (first and second quotes); Succession of Victorin Roman (and wife Celeste Judice), 16 Apr. 1832, Probate Records, SJPC (third and fourth quotes).

47. Succession of Margarete A. Hooke, widow of C. M. Shepherd, 14 Apr. 1850, Probate Records, SJPC (quote); Persac, "Plantations on the Mississippi River," LC.

48. Sitterson, *Sugar Country*, 196-204; Succession of Auguste Gaudet, 18 Apr. 1832, Probate Records, SJPC (first and second quotes); Succession of Jean Baptiste Chastant, 10 Jan. 1832 (third and fourth quotes).

49. Smith and Smith, *Cane, Cotton, and Crevasses*, 81; Russell, *My Diary*, 1:397 (first quote); *Le Messager*, 28 Apr. 1849 (second quote); *L' Avant Coureur*, 20 Jan. 1856 (third quote); *La Messager*, 8 Sept. 1852 (fourth quote); Clayton, *Mother Wit*, 166 (sixth quote), 179 (seventh quote). For other examples see also Olmsted, *A Journey*, 660; Pulszky, *White, Red, Black*, 2:103.

50. Pulszky, *White, Red, Black*, 2:102 (first quote); *Planter's Advocate*, 23 May 1840 (second quote); Succession of Jacques Telesphore Roman, 14 Apr. 1848, Probate Records, SJPC.

51. Succession of Joseph Dugas, 8 Feb. 1833, Probate Records, SJPC; Succession of Patrick Uriell, 23 Apr. 1835, Probate Records, SJPC; Valcour Aime Slave Records, 1821-1850, RASP, Series H, LSU. See also Malone, *Sweet Chariot*, 213.

52. Stephenson, *Isaac Franklin*, 77; Succession of Jean Baptiste Keller, 14 Jan. 1833, Probate Records, SJPC (quote); Succession of Zenon Arcenaux, 4 Aug. 1848, Probate Records, SJPC; Succession of Pierre Chenet, 28 Aug. 1837, Probate Records, SJPC.

53. Succession of Auguste Gaudet, 18 Apr. 1832, Probate Records, SJPC (first quote); *Le Messager*, 22 Jun. 1854; Clayton, *Mother Wit*, 162 (second quote), 179 (third quote).

54. Malone, *Sweet Chariot*, 212-13 (first quote); Succession of Patrick Uriell, 23 Apr. 1835, Probate Records, SJPC (third quote); Olmsted, *A Journey*, 678 (fourth quote).

55. Johnson, *Soul by Soul*, 1-2; succession of Jean Baptiste Boucry, 30 Sept. 1833, Probate Records, SJPC; Clayton, *Mother Wit*, 103 (first quote); *L'Avant Coureur*, 26 Feb. 1854 (second quote); *Louisiana State Gazette*, 13 Feb. 1826 (third quote).

56. Follett, *Sugar Masters*, 83, 85 (second quote), 86; Cash Book, 25 July 1854, 25 Dec. 1853, Bruce, Seddon and Wilkins Plantation Records, LSU; Jean Baptiste Ferchand, Journal, 1858, RASP, Series H, LSU; Sitterson, *Sugar Country*, 66-67. See also Northup, *Twelve Years*, 208.

57. E. Herbert to A. E. Crane, 6 Oct. 1858, Andrew E. Crane Family Papers, LSU (quote).

Chapter 8. Weathering Different Storms

1. For more on the transition from slavery to freedom in these regions and the South in general, see the following works, among others: Williamson, *After Slavery*; Ripley, *Slaves and Freedmen*; Litwack, *Been in the Storm So Long*; Foner, *Reconstruction*; Schwalm, *Hard Fight*; Berlin and Rowland, *Families and Freedom*; Rodrigue, *Reconstruction*.

2. Parish, *Slavery*, 97 (quote).

Bibliography

Primary Sources

Alexandria Public Library, Alexandria, Virginia

Fairfax County Deed Books, 1800–1860 (microfilm)

Charleston County Courthouse, Charleston, South Carolina

Charleston County Wills

Fairfax City Regional Library, Fairfax, Virginia

Fairfax County Will Books (microfilm)
Diary of Richard Marshall Scott, Sr. (typescript)
Diary of Richard Marshall Scott, Jr. (typescript)
Diary of Virginia Gunnell Scott (typescript)

Fisher Library, University of Virginia, Charlottesville, Virginia

Online Historical Census Browser. http://fisher.lib.virginia.edu/collections/stats/histcensus/index.html.

Georgetown County Courthouse, Georgetown, South Carolina

Georgetown County Wills

Gunston Hall Plantation Archival Collections, Lorton, Virginia

Graham Family Papers
Thomas Francis Mason Papers

Hill Memorial Library, Louisiana State University, Baton Rouge, Louisiana

Ashland Plantation Record Book
Boucry Family Record Books
Louis Bringier and Family Papers
Bruce, Seddon & Wilkins Plantation Records
Andrew E. Crane Family Papers

236 / Bibliography

Keller Family Plantation Records
Severin Landry Papers
George Mather Account Books
Rosalie Andry Palao, Bill of Sale
Records of Ante-Bellum Southern Plantations: From the Revolution through the Civil War, edited by Kenneth M. Stampp (microfilm series).
H. M. Seale Diary
Benjamin Tureaud Family Papers
Uncle Sam Plantation Papers
Welham Plantation Record Books
William Webb Wilkins Papers

Howard-Tilton Memorial Library, Tulane University, New Orleans

Octave Colomb Plantation Journal
Jean Baptiste Ferchand Journal
Eugene Forstall Letterbooks

Library of Congress, Washington, D.C.

Custis-Lee Family Papers
Stephen D. Doar Papers
Edward Frost Papers
William Lowndes Papers
Bushrod Washington Papers
David Wilson Scott Papers
Hoffman, J. Paul. "A map of Fairfax County, and parts of Loudoun and Prince William Counties, Va., and the District of Columbia," Topographical Office, Army of Northern Virginia, 29 March 1864.
Mills, Robert. "Atlas of the State of South Carolina, 1825."
Persac, A. "Plantations on the Mississippi River from Natchez to New Orleans, 1858."

Library of Virginia, Richmond, Virginia

Legislative Petitions to the General Assembly of Virginia

National Archives and Records Administration, Washington, D.C.

Records of the Field Offices for the State of Louisiana, Bureau of Refugees, Freedmen, and Abandoned Lands, 1861–1869 [RG 105]. Marriage Records of the Office of the Commissioner.
———. Register of Black Persons, Subordinate Office, Donaldsonville, La.
———. Ross Home Colony, Register of Arrivals and Departures, Feb. 1865-Jul. 1866.
U.S. Bureau of the Census. Population Census of the United States, 1800–1860 (microfilm).
———. Nonpopulation Census Schedules, Agriculture, 1850–1860 (microfilm)

Robert W. Woodruff Library, Emory University, Atlanta, Georgia

Keith M. Read Confederate Collection

South Carolina Historical Society, Charleston, South Carolina

R. F. W. Allston Papers
Cheves-Middleton Papers
Gourdin-Gaillard Family Papers
Pringle Family Papers
Joshua John Ward Plantation Journals

South Caroliniana Library, University of South Carolina, Columbia, South Carolina

Cleland Kinloch Huger Papers
Miles, C. R. "In equity, Charleston District: Emily Frances Weston, executrix of Plowden C. J. Weston, deceased . . ." 1864.
Davison McDowell Papers
Read-Lance Family Papers
James Ritchie Sparkman Papers

St. James Parish Courthouse, Convent, Louisiana

St. James Parish Probate Records, 1800–1860

Secondary Sources

Abbott, Richard H. "Yankee Farmers in Northern Virginia, 1840–1860." *Virginia Magazine of History and Biography* 76 (Jan. 1968): 56–63.

Abdy, E. S. *Journal of a Residence and Tour in the United States of North America from April 1833 to October 1834*. 3 vols. First published 1835; reprint New York: Negro Universities Press, 1969.

Aime, Valcour. *Plantation Diary of the Late Mr. Valcour Aime, Formerly the Proprietor of the Plantation Known as the St. James Sugar Refinery, Situated in the Parish of St. James, and Now Owned by Mr. John Burnside*. New Orleans: Clark & Hofeline, 1878.

Albert, Octavia V. Rogers. *The House of Bondage, or Charlotte Brooks and Other Slaves*. First published 1890; reprint New York: Oxford University Press, 1988.

Allen, Alexander V. G. *Life and Letters of Phillips Brooks*. 3 vols. New York: E. P. Dutton, 1901.

Allston, Elizabeth Deas. *The Allstons and Alstons of Waccamaw*. Charleston, S.C.: Walker, Evans, & Cogswell, 1936.

Alston, J. Motte. *Rice Planter and Sportsman: The Recollections of J. Motte Alston, 1821–1909*, edited by Arney R. Childs. First published 1953; reprint Columbia, S.C.: University of South Carolina Press, 1999.

Andrews, Ethan Allen. *Slavery and the Domestic Slave Trade in the United States, in a*

Series of Letters Addressed to the Executive Committee of the American Union for the Relief and Improvement of the Colored Race. First published 1836; reprint Freeport, N.Y.: Books for Libraries Press, 1971.

Anonymous. "A Slave's Story." *Putnam's Monthly Magazine* 9 (June 1857): 614–20.

Bailey, David Thomas. "A Divided Prism: Two Sources of Black Testimony on Slavery," *Journal of Southern History* 46 (Aug. 1980): 381–404.

Bancroft, Frederic. *Slave-Trading in the Old South.* First published 1931; reprint Columbia, S.C.: University of South Carolina Press, 1996.

Bauer, Craig A. *A Leader among Peers: The Life and Times of Duncan Farrar Kenner.* Lafayette: Center for Louisiana Studies, University of Southwestern Louisiana, 1993.

Begnaud, Allen. "The Louisiana Sugar Cane Industry: An Overview." In *Green Fields, Green Fields: Two Hundred Years of Louisiana Sugar,* compiled by the Center for Louisiana Studies, University of Southwestern Louisiana, 29–50. Lafayette, La.: The Center, 1980.

Berlin, Ira. *Generations of Captivity: A History of African-American Slaves.* Cambridge, Mass.: Harvard University Press, 2003.

———. *Many Thousands Gone: The First Two Centuries of Slavery in North America.* Cambridge, Mass.: Harvard University Press, 1998.

Berlin, Ira and Leslie Rowland, eds. *Families and Freedom: A Documentary History of African-American Kinship in the Civil War Era.* New York: New Press, 1997.

Berry, Daina Ramey. *Swing the Sickle for the Harvest Is Ripe: Gender and Slavery in Antebellum Georgia.* Urbana: University of Illinois Press, 2007.

Blassingame, John W. *The Slave Community: Plantation Life in the Antebellum South.* First published 1972; reprint New York: Oxford University Press, 1979.

———. "Using the Testimony of Ex-Slaves: Approaches and Problems." *Journal of Southern History* 41 (Nov. 1975): 473–92.

———, ed. *Slave Testimony: Two Centuries of Letters, Speeches, Interviews, and Autobiographies.* Baton Rouge: Louisiana University Press, 1977.

Bolland, O. Nigel "Proto-Proletarians? Slave Wages in the Americas." In *From Chattel Slaves to Wage Slaves: The Dynamics of Labour Bargaining in the Americas,* edited by Mary Turner, 123–47. Bloomington: University of Indiana Press, 1995.

Bourgeois, Lillian C. *Cabanocey: The History, Customs and Folklore of St. James Parish.* New Orleans: Pelican, 1957.

Boyle, Christopher and James A. Fitch, eds. *Georgetown County Slave Narratives.* Georgetown, S.C.: Rice Museum, 1997.

Brady, Patricia, ed. *George Washington's Beautiful Nelly: The Letters of Eleanor Parke Custis Lewis to Elizabeth Bordley Gibson, 1794–1851.* Columbia: University of South Carolina Press, 1991.

Breen, T. H. *Tobacco Culture: The Mentality of the Great Tidewater Planters on the Eve of Revolution.* Princeton, N.J.: Princeton University Press, 1985.

Bremer, Fredrika. *Homes of the New World; Impressions of America.* 2 vols. New York: Harper & Row, 1854.

Briesen, Martha von, ed. *Letters of Elijah Fletcher*. Charlottesville: University of Virginia Press, 1965.
Buckingham, J. S. *The Slave States of America*. 2 vols. London: Fisher, 1842.
Bull, Henry DeSaussure. *All Saints' Church, Waccamaw: The Parish, the Place, the People, 1739–1948*. Columbia, S.C.: R. L. Bryan, 1949.
Camp, Stephanie M. H. *Closer to Freedom: Enslaved Women and Everyday Resistance in the Plantation South*. Chapel Hill: University of North Carolina Press, 2004.
Carney, Judith Ann. *Black Rice: The African Origins of Rice Cultivation in the Americas*. Cambridge, Mass.: Harvard University Press, 2001.
Cashin, Joan E. "Landscape and Memory in Antebellum Virginia." *Virginia Magazine of History and Biography* 102 (Oct. 1994): 477–500.
Champomier, P. A. *Statement of the Sugar Crop in Louisiana*. New Orleans: Magne & Weisse, 1849–60.
Chaplin, Joyce E. "Tidal Rice Cultivation and the Problem of Slavery in South Carolina and Georgia, 1760–1815." *William and Mary Quarterly* 49 (Jan. 1992): 29–61.
Clayton, Ronnie W. *Mother Wit: The Ex-Slave Narratives of the Louisiana Writers' Project*. New York: P. Lang, 1990.
Coclanis, Peter A. "How the Low Country Was Taken to Task: Slave-Labor Organization in Coastal South Carolina and Georgia." In *Slavery, Secession, and Southern History*, edited by Robert Louis Paquette and Louis A. Ferleger, 59–80. Charlottesville: University of Virginia Press, 2000.
———. "Rice Prices in the 1720s and the Evolution of the South Carolina Economy." *Journal of Southern History* 48 (Nov. 1982): 531–44.
Collins, Elizabeth. *Memories of the Southern States*. Taunton, England: Barnicott, 1865.
Conway, M. D. *Testimonies Concerning Slavery*. First published 1865; reprint New York: Arno Press, 1969.
Craven, Avery Odell. *Soil Exhaustion as a Factor in the Agricultural History of Virginia and Maryland, 1606–1860*. Urbana: University of Illinois Press, 1926.
Crawford, Stephen C. "Quantitative Memory: A Study of the WPA and Fisk University Slave Narrative Collections." PhD diss., University of Chicago, 1980.
Crété, Lillian. *Daily Life in Louisiana, 1815–1830*. Baton Rouge: Louisiana State University Press, 1978.
David, Paul A., et al. *Reckoning with Slavery: A Critical Study in the Quantitative History of American Negro Slavery*. New York: Oxford University Press, 1976.
Davis, David Brion. "A Review of the Conflicting Theories on the Slave Family." *Journal of Blacks in Higher Education* 16 (Summer 1997): 100–103.
Davis, John. *Travels of Four Years and a Half in the United States of America during 1798, 1799, 1800, 1801, and 1802*. First published 1803; reprint New York: H. Holt, 1909.
Devereux, Anthony Q. *The Rice Princes: An Epoch Revisited*. Columbia, S.C.: State Printing Co., 1973.
Deyle, Steven. *Carry Me Back: The Domestic Slave Trade in American Life*. New York: Oxford University Press, 2005.

Din, Gilbert C. *Spaniards, Planters, and Slaves: The Spanish Regulation of Slavery in Louisiana, 1763-1803*. College Station: Texas A&M University Press, 1999.

Doar, David. *Rice and Rice Planting in the South Carolina Low Country*. Charleston: Charleston Museum, 1936.

Drago, Edmund L., ed. *Broke by the War: Letters of a Slave Trader*. Columbia: University of South Carolina Press, 1991.

Drake, Pam. *Pictures of the "Peculiar Institution" as It Exists in Louisiana and Mississippi, by an Eye-Witness*. Boston: J. B. Yerrington, 1850.

Drew, Benjamin. *A North-Side View of Slavery: The Refugee, or the Narratives of Fugitive Slaves in Canada*. First published 1856; reprint New York: Negro Universities Press, 1968.

Du Bois, W. E. B. *The Negro American Family*. First published in 1909; reprint Cambridge, Mass.: M.I.T. Press, 1970.

Dunaway, Wilma. *The African-American Family in Slavery and Emancipation*. New York: Cambridge University Press, 2003.

Dusinberre, William. *Them Dark Days: Slavery in the American Rice Swamps*. New York: Oxford University Press, 1996.

Dziobek, Linda A., et al. "Mount Erin: A Bit of Old Sod in Fairfax County." *Yearbook of the Historical Society of Fairfax County, Virginia* 24 (1993-1994): 28-55.

Easterby, J. H., ed. *The South Carolina Rice Plantation as Revealed in the Papers of Robert F. W. Allston*. Chicago: University of Chicago Press, 1945.

Elkins, Stanley M. *Slavery: A Problem in American Institutional and Intellectual Life*. Chicago: University of Chicago Press, 1959.

Evans, Estwick. *A Pedestrious Tour of Four Thousand Miles through the Western States and Territories during the Winter and Spring of 1818*. Concord, N.H.: Joseph C. Spear, 1819.

Federal Writers' Project. *Slave Narratives: A Folk History of the United States of America from Interviews with Former Slaves, 1936-1938*. 17 vols. Washington, D.C., 1941.

Fogel, Robert William. *Without Consent or Contract: The Rise and Fall of American Slavery*. New York: W. W. Norton, 1989.

Fogel, Robert William and Stanley L. Engerman, *Time on the Cross: The Economics of American Negro Slavery*. Boston: Little, Brown, 1974.

Follett, Richard. *The Sugar Masters: Planters and Slaves in Louisiana's Cane World, 1820-1860*. Baton Rouge: Louisiana State University Press, 2005.

Foner, Eric. *Reconstruction: America's Unfinished Revolution, 1863-1877*. New York: Harper & Row, 1988.

Foshee, Andrew W. "Slave Hiring in Rural Louisiana." *Louisiana History* 26 (Winter 1985): 63-73.

Franklin, John Hope and Loren Schweninger. *Runaway Slaves: Rebels on the Plantation*. New York: Oxford University Press, 1999.

Frazier, E. Franklin. *The Negro Family: A Study of Family Origins before the Civil War*. Nashville, Tenn.: Fisk University Press, 1932.

Frobel, Anne S. *The Civil War Diary of Anne S. Frobel of Wilton Hill in Virginia*. Florence, Ala.: M. H. and D. M. Lancaster, 1986.

Gamble, Robert S. *Sully: The Biography of a House.* Chantilly, Va.: Sully Foundation, 1973.
Gayarré, Charles. "A Louisiana Sugar Plantation of the Old Régime," *Harper's New Monthly Magazine* 74 (1886–87): 606–22.
Genovese, Eugene D. *Roll, Jordan, Roll: The World the Slaves Made.* New York: Pantheon Books, 1974.
Gill, Harold B., Jr., "Wheat Culture in Colonial Virginia." *Agricultural History* 52 (July 1978): 380–93.
Gore, Laura Locoul. *Memories of the Old Plantation Home.* Vacherie, La.: Zoë Company, 2001.
Gourdin, J. Raymond. *Gourdin: The History and Genealogy of a French-American Family from Georgetown County, South Carolina, 1830–1994.* Baltimore: Gateway Press, 1995.
Gragg, Rod. *Pirates, Planters, and Patriots: Historical Tales from the South Carolina Grand Strand.* Winston-Salem, N.C.: Peace Hill, 1984.
Gray, Lewis Cecil. *History of Agriculture in the Southern United States to 1860.* 2 vols. Washington, D.C.: Carnegie Institution of Washington, 1933.
Gutheim, Frederick. *The Potomac.* New York: Rinehart, 1949.
Gutman, Herbert G. *The Black Family in Slavery and Freedom, 1750–1925.* New York: Vintage, 1976.
———. *Slavery and the Numbers Game: A Critique of Time on the Cross.* First published in 1975; reprint Urbana: University of Illinois Press, 2003.
Hamilton, Capt. Thomas. *Men and Manners in America.* 2 vols. Edinburgh: W. Blackwood, 1834.
Hendrix, James Paisley, Jr. "The Efforts to Reopen the African Slave Trade in Louisiana." *Louisiana History* 10 (Spring 1969): 97–123.
Henry, H. M. *The Police Control of the Slave in South Carolina.* Emory, Va., 1914.
Hodgson, Adam. *Remarks during a Journey through North America in the Years 1819, 1820, and 1821, in a Series of Letters.* First published 1823; reprint Westport, Conn.: Negro Universities Press, 1970.
Hudson, Larry E., Jr. "'All that Cash': Work and Status in the Slave Quarters." In *Working toward Freedom: Slave Society and Domestic Economy in the American South*, edited by Larry E. Hudson, Jr., 77–94. Rochester, N.Y.: University of Rochester Press, 1994.
———. *To Have and to Hold: Slave Work and Family Life in Antebellum South Carolina.* Athens: University of Georgia Press, 1997.
Hurst, Harold W. *Alexandria on the Potomac: The Portrait of an Antebellum Community.* Lanham, Md.: University Press of America, 1991.
Ingraham, Joseph Holt. *The South-West, by a Yankee.* 2 vols. New York: Harper & Bros., 1835.
Irwin, James R. "Exploring the Affinity of Wheat and Slavery in the Virginia Piedmont." *Explorations in Economic History* 25 (1988): 295–322.
Janney, Samuel McPherson. *Memoirs of Samuel M. Janney.* Philadelphia, 1881.
Janney, Werner L. and Asa Moore Janney, eds. *John Jay Janney's Virginia: An American Farm Lad's Life in the Early 19th Century.* McLean, Va.: EPA Publications, 1978.

Johnson, Walter. *Soul by Soul: Life inside the Antebellum Slave Market*. Cambridge, Mass.: Harvard University Press, 1999.

Johnston, James Hugo. *Race Relations in Virginia and Miscegenation in the South, 1776–1860*. First published 1937; reprint Amherst: University of Massachusetts Press, 1970.

Johnston, J. S. *Letter by Mr. Johnston of Louisiana to the Secretary of the Treasury, in Reply to His Circular of the 1st July, 1830, Relative to the Culture of Sugar Cane*. Washington, D.C.: Gales & Seaton, 1831.

Jones, Jacqueline. *Labor of Love, Labor of Sorrow: Black Women, Work, and the Family from Slavery to the Present*. New York: Basic Books, 1985.

Joyner, Charles. *Down by the Riverside: A South Carolina Slave Community*. Urbana: University of Illinois Press, 1984.

King, Wilma, ed. *A Northern Woman in the Plantation South: Letters of Tryphena Blanche Holder Fox, 1856–1876*. Columbia: University of South Carolina Press, 1993.

———. *Stolen Childhood: Slave Youth in Nineteenth-Century America*. Bloomington: Indiana University Press, 1995.

Kingsford, William. *Impressions of the West and South during a Six Weeks' Holiday*. Toronto: A. H. Armour, 1858.

Klingaman, David. "The Significance of Grain in the Development of the Tobacco Colonies," *Journal of Economic History* 29 (June 1969): 268–78.

Koger, Larry. *Black Slaveowners: Free Black Slave Masters in South Carolina, 1790–1860*. Columbia: University of South Carolina Press, 1994.

Kolchin, Peter. *American Slavery, 1619–1877*. New York: Hill & Wang, 1993.

Kulikoff, Allan. *Tobacco and Slaves: The Development of Southern Cultures in the Chesapeake, 1680–1800*. Chapel Hill: University of North Carolina Press, 1986.

Lachicotte, Alberta Morel. *Georgetown Rice Plantations*. First published 1955; reprint Georgetown, S.C.: Georgetown County Historical Society, 1993.

Le Gardeur, René J., Jr. "The Origins of the Sugar Industry in Louisiana." In *Green Fields: Two Hundred Years of Louisiana Sugar*, compiled by the Center for Louisiana Studies, University of Southwestern Louisiana, 1–28. Lafayette, La.: Center, 1980.

Leon, J. A. *On Sugar Cultivation; in Louisiana, Cuba, &c. and the British Possessions*. London: J. Ollivier, 1848.

Littlefield, Daniel. *Rice and Slaves: Ethnicity and the Slave Trade in Colonial South Carolina*. Baton Rouge: Louisiana State University Press, 1981.

Litwack, Leon F. *Been in the Storm So Long: The Aftermath of Slavery*. New York: Knopf, 1979.

Lyell, Charles. *A Second Visit to the United States of North America*. New York: Harper & Bros., 1850.

Machen, Arthur W., Jr. *Letters of Arthur W. Machen with Biographical Sketch*. Baltimore: s.n., 1917.

Malet, William Wyndham. *An Errand to the South in the Summer of 1862*. London: R. Bentley, 1863.

Malone, Ann Patton. *Sweet Chariot: Slave Family and Household Structure in Nineteenth-Century Louisiana*. Chapel Hill: University of North Carolina Press, 1992.
Martin, Jonathan D. *Divided Mastery: Slave Hiring in the American South*. Cambridge, Mass.: Harvard University Press, 2004.
Martineau, Harriet. *Retrospect of Western Travel*. 2 vols. First published 1838; reprint New York: Haskell House, 1969.
McDonald, Roderick A. *The Economy and Material Culture of Slaves: Goods and Chattels on the Sugar Plantations of Jamaica and Louisiana*. Baton Rouge: Louisiana State University Press, 1993.
———. "Independent Economic Production by Slaves on Antebellum Louisiana Sugar Plantations." In *Cultivation and Culture: Labor and the Shaping of Slave Life in the Americas*, edited by Ira Berlin and Philip D. Morgan, 182–208. Charlottesville: University of Virginia Press, 1993.
McMillen, Sally G. *Southern Women: Black and White in the Old South*. Wheeling, Ill.: Harlan Davidson, 2002.
McMillon, Lynn and Jane K. Wall, eds. *Fairfax County, Virginia: 1820 Federal Population Census and Census of Manufactures*. Vienna, Va.: McMillon, 1976.
McPherson, Samuel. *Memories of Samuel M. Janney*. Philadelphia: s.n., 1881.
Meaders, Daniel, ed. *Advertisements for Runaway Slaves in Virginia, 1800–1820*. New York: Garland, 1997.
Menn, Joseph Karl. *The Large Slaveholders of Louisiana in 1860*. New Orleans: Pelican, 1964.
Michie, James L. *Richmond Hill Plantation, 1810–1868: The Discovery of Antebellum Life on a Waccamaw Rice Plantation*. Spartanburg, S.C.: Reprint Co., 1990.
———. *Richmond Hill and Wachesaw: An Archaeological Study of Two Rice Plantations on the Waccamaw River, Georgetown County, South Carolina*. Columbia, S.C.: The Institute, 1987.
Mills, Robert. *Statistics of South Carolina*. Charleston, S.C.: Hurlbut & Lloyd, 1826.
Mintz, Sidney W. "The Question of Caribbean Peasantries: A Comment," *Caribbean Studies* 1 (Oct. 1961): 31–34.
Moody, V. Alton. *Slavery on Louisiana Sugar Plantations*. New Orleans: Cabildo, 1924.
Morgan, Kenneth. "Slave Sales in Colonial Charleston." *English Historical Review* 113 (Sept. 1998): 905–27.
Morgan, Philip D. "The Ownership of Property by Slaves in the Mid-Nineteenth-Century Low Country." *Journal of Southern History* 49 (Aug. 1983): 399–420.
———. *Slave Counterpoint: Black Culture in the Eighteenth-Century Chesapeake & Lowcountry*. Chapel Hill: University of North Carolina Press, 1998.
———. "Work and Culture: The Task System and the World of Lowcountry Blacks, 1700–1880." *William and Mary Quarterly* 39 (Oct. 1982): 563–99.
Moynihan, Daniel P. "The Negro Family: The Case for National Action." Washington, D.C.: Office of Policy Planning Research, United States Department of Labor, 1965.
Muir, Dorothy Troth. *Potomac Interlude: The Story of Woodlawn Mansion and the Mount Vernon Neighborhood, 1846–1943*. Washington, D.C.: Mount Vernon, 1943.

Netherton, Nan, et al. *Fairfax County, Virginia: A History*. Fairfax, Va.: Fairfax County Board of Supervisors, 1978.

Netherton, Ross and Nan. *Green Spring Farm, Fairfax County, Virginia*. Fairfax, Va.: Division of Planning, 1978.

Nevins, Allan, ed. *American Social History as Recorded by British Travellers*. New York: H. Holt, 1923.

Niccolls, Jeanne. "Old John: In Search of His Story." *Yearbook of the Historical Society of Fairfax County, Virginia* 27 (1999–2000): 71–80.

Niemcewicz, Julian Ursyn. *Under Their Vine and Fig Tree: Travels through America in 1797–1799, 1805 with some Further Account of Life in New Jersey*, translated and edited by Metchie J. E. Budka. Elizabeth, N.J.: Grassmann, 1965.

Northup, Solomon. *Twelve Years a Slave*. First published 1853; reprint Mineola, N.Y.: Dover, 1970.

Oliphant, Laurence. *Patriots and Filibusters; or, Incidents of Political and Exploratory Travel*. London: W. Blackwood, 1860.

Olmsted, Frederick Law. *A Journey in the Seaboard Slave States in the Years 1853–1854, with Remarks on their Economy*. New York: Mason Brothers, 1861.

Owens, Leslie Howard. *This Species of Property: Slave Life and Culture in the Old South*. New York: Oxford University Press, 1976.

Pargas, Damian Alan. "Boundaries and Opportunities: Comparing Slave Family Formation in the Antebellum South." *Journal of Family History* 33 (July 2008): 316–45.

———. "Disposing of Human Property: American Slave Families and Forced Separation in Comparative Perspective," *Journal of Family History* 34 (July 2009): 251–74.

———. "'Various Means of Providing for Their Own Tables': Comparing Slave Family Economies in the Antebellum South." *American Nineteenth Century History* 7 (Sept. 2006): 361–87. http://www.informaworld.com.

———. "Work and Slave Family Life in Antebellum Northern Virginia." *Journal of Family History* 31 (Oct. 2006): 335–57.

Parish, Peter J. *Slavery: History and Historians*. New York: Harper & Row, 1989.

Parker, Amos A. *Trip to the West and Texas*. Boston: Benjamin B. Mursey, 1836.

Perdue, Charles, Jr., et al, eds. *Weevils in the Wheat: Interviews with Virginia Ex-Slaves*. Charlottesville: University of Virginia Press, 1976.

Peterson, Arthur G. "The Alexandria Market Prior to the Civil War." *William and Mary Quarterly* 12 (Apr. 1932): 104–14.

Phillips, Ulrich Bonnell. *American Negro Slavery: A Survey of the Supply, Employment and Control of Negro Labor as Determined by the Plantation Regime*. New York: Appleton, 1918.

———. *Plantation and Frontier, 1649–1863*. 2 vols. New York: B. Franklin, 1969.

———. *The Slave Economy of the Old South: Selected Essays in Economic and Social History*, edited by Eugene D. Genovese. Baton Rouge: Louisiana State University Press, 1968.

Poland, Charles P., Jr. *From Frontier to Suburbia*. Marceline, Mo.: Walsworth, 1976.

Pritchard, Walter. "Routine on a Louisiana Sugar Plantation under the Slavery Régime," *Mississippi Valley Historical Review* 14 (Sept. 1927): 168–78.

———, ed. "A Tourist's Description of Louisiana in 1860." *Louisiana Historical Quarterly* 21 (Oct. 1938): 11–15.
Pringle, Elizabeth W. Allston. *Chronicles of Chicora Wood*. First published 1922; reprint Atlanta: Cherokee, 1976.
Pryor, Elizabeth Brown. "Flexibility and Profit in the Slave Hiring System in Fairfax County, Virginia, 1830–1860." Unpublished article, 1980.
———. *Walney: Two Centuries of a Northern Virginia Plantation*. Fairfax, Va.: History and Archaeology Section, Office of Comprehensive Planning, 1984.
Pulszky, Francis and Theresa. *White, Red, Black*. First published 1853; reprint New York: Johnson Reprint Co., 1970.
Rafuse, Diane N. *Maplewood*. Fairfax, Va.: Division of Planning, 1970.
Redpath, James. *The Roving Editor, or Talks with Slaves in the Southern States*. First published 1859; reprint University Park, Penn.: Pennsylvania State University Press, 1996.
Ricards, Sherman L., and George M. Blackburn. "A Demographic History of Slavery: Georgetown County, South Carolina, 1850." *South Carolina Historical Magazine* 76 (Oct. 1975): 215–24.
Ripley, C. Peter. *Slaves and Freedmen in Civil War Louisiana*. Baton Rouge: Louisiana State University Press, 1976.
Ripley, Eliza. *Social Life in Old New Orleans, being Recollections of My Childhood*. New York: D. Appleton, 1912.
Rhyne, Nancy, ed. *Voices of South Carolina Slave Children*. Orangeburg, S.C.: Sandlapper, 1999.
Robert, Joseph C. "Lee the Farmer." *Journal of Southern History* 3 (Nov. 1937): 422–40.
Robinson, Solon. *Solon Robinson, Pioneer and Agriculturalist: Selected Writings*. Edited by Herbert Anthony Kellar. 2 vols. Indianapolis: Indiana Historical Bureau, 1936.
Rodrigue, John C. *Reconstruction in the Cane Fields: From Slavery to Free Labor in Louisiana's Sugar Parishes*. Baton Rouge: Louisiana State University Press, 2001.
Rogers, George C., Jr. *The History of Georgetown County, South Carolina*. First published 1970; reprint Spartanburg, S.C.: Published for the Georgetown Historical Society by Reprint Publishing Co., 1990.
Rowland, Dunbar, ed. *Official Letter Books of W. C. C. Claiborne, 1801–1816*. 6 vols. Jackson, Miss.: State Department of Archives and History, 1917.
Russell, William Howard, *My Diary North and South*. 2 vols. London: Bradbury & Evans, 1863.
Sallient, John, ed. *Afro-Virginian History and Culture*. New York: Garland, 1999.
Savitt, Todd L. *Medicine and Slavery: The Diseases and Health Care of Blacks in Antebellum Virginia*. Urbana: University of Illinois Press, 1978.
Schlotterbeck John T. "The Internal Economy of Slavery in Rural Piedmont Virginia" In *The Slaves' Economy: Independent Production by Slaves in the Americas*, edited by Ira Berlin and Philip D. Morgan, 170–81. London: Frank Cass, 1991.
Schmitz, Mark D. "Economies of Scale and Farm Size in the Antebellum Sugar Sector." *Journal of Economic History* 37 (Dec. 1977): 959–80.
Schwaab, Eugene L., ed. *Travels in the Old South, Selected from the Periodicals of the Times*. 2 vols. Lexington: University Press of Kentucky, 1973.

Schwalm, Leslie A. *A Hard Fight for We: Women's Transition from Slavery to Freedom in South Carolina*. Urbana: University of Illinois Press, 1997.

Sitterson, J. Carlyle. "Magnolia Plantation, 1852–1862: A Decade of a Louisiana Sugar Estate." *Mississippi Valley Historical Review* 25 (Sept. 1938): 197–210.

———. "The McCollams: A Planter Family of the Old and New South." *Journal of Southern History* 6 (Aug. 1940): 347–67.

———. *Sugar Country: The Cane Sugar Industry in the South, 1753–1950*. Lexington: University Press of Kentucky, 1953.

———. "The William J. Minor Plantations: A Study in Ante-Bellum Absentee Ownership." *Journal of Southern History* 9 (Feb. 1943): 59–74.

Smith, Eugenia B. *Centreville, Virginia: Its History and Architecture*. Fairfax, Va.: Fairfax County Office of Planning, 1973.

Smith, James L. *Autobiography of James L. Smith*. First published 1881; reprint Miami: Mnmemosyne, 1969.

Smith, Mark M. *Debating Slavery: Economy and Society in the Antebellum South*. Cambridge: Cambridge University Press, 1998.

Smith, Philip Chadwick Foster and G. Gouverneur Meredith S. Smith. *Cane, Cotton, and Crevasses: Some Antebellum Louisiana and Mississippi Plantations of the Minor, Kenner, Hooke, and Shepherd Families*. Bath, Maine: Renfrew Group, 1992.

Spann, Barbara J. *Carlby*. Fairfax, Va.: Fairfax County Office of Comprehensive Planning, 1976.

Spindel, Donna J. "Assessing Memory: Twentieth-Century Slave Narratives Reconsidered." *Journal of Interdisciplinary History* 27 (Autumn 1996): 247–61.

Sprouse, Edith Moore. *Mount Air, Fairfax County, Virginia*. Fairfax, Va.: Division of Planning, 1970.

Stampp, Kenneth M. *The Peculiar Institution: Slavery in the Ante-bellum South*. New York: Knopf, 1956.

Stephenson, Wendell Holmes. *Isaac Franklin: Slave Trader and Planter of the Old South*. First published 1938; reprint Gloucester, Mass.: P. Smith, 1968.

Sterling, Dorothy, ed. *We Are Your Sisters: Black Women in the Nineteenth Century*. New York: W. W. Norton, 1984.

Sternberg, Mary Ann. *Along the River Road: Past and Present on Louisiana's Historic Byway*. Baton Rouge: Louisiana State University Press, 2001.

Stevenson, Brenda E. *Life in Black & White: Family and Community in the Slave South*. New York: Oxford University Press, 1996.

Steward, Austin. *Twenty-Two Years a Slave and Forty Years a Freeman*. First published 1857; reprint Syracuse, N.Y.: Syracuse University Press, 2002.

Still, William. *The Underground Railroad: A Record of Facts, Authentic Letters, etc.* First published 1872; reprint New York: Arno Press, 1968.

Stirling, James. *Letters from the Slave States*. First published 1857; reprint New York: Negro Universities Press, 1969.

Stuart, James. *Three Years in North America*. 2 vols. Edinburgh: R. Cadell, 1833.

Stuckey, Sterling. *Slave Culture: Nationalist Theory and the Foundations of Black America*. New York: Oxford University Press, 1987.

Sutch, Richard. "The Breeding of Slaves for Sale and the Westward Expansion of Slavery, 1850–1860." In *Race and Slavery in the Western Hemisphere: Quantitative Studies*., edited by Stanley Engerman and Eugene D. Genovese, 173–210. Princeton, N.J.: Princeton University Press, 1975.

Sutcliffe, Robert. *Travels in Some Parts of North America in the Years 1804, 1805, and 1806*. Philadelphia: B & T Kite, 1812.

Swan, Dale E. "The Structure and Profitability of the Antebellum Rice Industry: 1859." *Journal of Economic History* 31 (Mar. 1973): 321–25.

Sweig, Donald M. "The Importation of African Slaves to the Potomac River, 1732–1772." *William & Mary Quarterly* 42 (Oct. 1985): 507–24.

———. "Northern Virginia Slavery: A Statistical and Demographic Investigation." PhD diss., College of William and Mary, 1982.

———. *Slavery in Fairfax County, Virginia, 1750–1860: A Research Report*. Fairfax, Va.: History and Archaeology Section, Office of Comprehensive Planning, 1983.

———., ed. *Registration of Free Negroes Commencing September Court 1822, Book No. 2, and Register of Free Blacks 1835, Book No. 3, Being the Full Text of the Two Extant Volumes, 1822–1861, of Registrations of Free Blacks Now in the County Courthouse, Fairfax, Virginia*. Fairfax, Va.: History Section, Office of Comprehensive Planning, 1977.

Tadman, Michael. "The Demographic Cost of Sugar: Debates on Slave Societies and Natural Increase in the Americas." *American Historical Review* 105 (Dec. 2000): 1534–75.

———. *Speculators and Slaves: Masters, Traders, and Slaves in the Old South*. Madison: University of Wisconsin Press, 1996.

Taylor, R. H. "Feeding Slaves." *Journal of Negro History* 9 (Apr. 1924): 139–43.

Taylor, Rosser H. *Ante-bellum South Carolina: A Social and Cultural History*. New York: Da Capo Press, 1970.

Teel, Dorothy O. *The 1810 Census of Georgetown District, South Carolina*. Hemingway, S.C.: Three Rivers Historical Society, 1995.

Tixier, Victor. *Tixier's Travels on the Osage Prairies*, edited by John Francis McDermott. First published 1844; reprint Norman: University of Oklahoma Press, 1940.

Thorpe, T. B. "Sugar and the Sugar Region of Louisiana." *Harper's New Monthly Magazine* 42 (Nov. 1853): 746–67.

Toledano, Roulhac. "Louisiana's Golden Age: Valcour Aime in St. James Parish." *Louisiana History* 10 (Summer 1969): 211–24.

Tower, Philo. *Slavery Unmasked: Being a Truthful Narrative of Three Years' Residence and Journeying in Eleven Southern States*. First published 1856; reprint New York: Negro Universities Press, 1969.

Trinkley, Michael, ed. *An Archaeological Study of Willbrook, Oatland, and Turkey Hill Plantations, Waccamaw Neck, Georgetown County, South Carolina*. Columbia, S.C.: Chicora Foundation, 1987.

Van Deburg, William L. *The Slave Drivers: Black Agricultural Labor Supervisors in the Antebellum South*. Westport, Conn.: Greenwood Press, 1979.

Veney, Bethany. *The Narrative of Bethany Veney, a Slave Woman*. Worcester, Mass.: s.n., 1889.

Virginian (Anon.). *The Yankees in Fairfax County*. Baltimore: Snodgrass & Wehrly, 1845.

Walsh, Lorena S. "Plantation Management in the Chesapeake, 1620–1820." *Journal of Economic History* 49 (June 1989): 393–406.

Weir, Robert M. *Colonial South Carolina: A History*. Millwood, N.Y.: KTO Press, 1983.

West, Emily. *Chains of Love: Slave Couples in Antebellum South Carolina*. Urbana: University of Illinois Press, 2004.

White, Deborah Gray. *Ar'n't I a Woman? Female Slaves in the Plantation South*. New York: Norton, 1985.

Whitten, David O. *Andrew Durnford: A Black Sugar Planter in the Antebellum South*. First published 1981; reprint New Brunswick, N.J.: Transaction, 1995.

Williamson, Joel. *After Slavery: The Negro in South Carolina during Reconstruction, 1861–1877*. Chapel Hill: University of North Carolina Press, 1965.

Wingfield, Charles L. "The Sugar Plantations of William J. Minor, 1830–1860." M.A. thesis, Louisiana State University, 1950.

Wood, Betty. *Women's Work, Men's Work: The Informal Slave Economies of Lowcountry Georgia*. Athens: University of Georgia Press, 1995.

Wood, Peter. *Black Majority: Negroes in Colonial South Carolina from 1670 through the Stono Rebellion*. New York: W. W. Norton, 1974.

Woodward, C. Vann. "History from Slave Sources." *American Historical Review* 79 (Apr. 1974): 470–81.

Workers of the Writers' Program of the Works Projects Administration in the State of Virginia. *The Negro in Virginia*. First published 1940; reprint Winston-Salem, N.C.: Blair, 1994.

Wren, Tony P. *Huntley: A Mason Family Country House*. Fairfax, Va.: Division of Planning, 1971.

Yakubik, Jill-Karen and Rosalinda Méndez. *Beyond the Great House: Archaeology at Ashland-Belle Helene Plantation*. Baton Rouge: Louisiana Dept. of Culture, Recreation, and Tourism, Division of Archaeology, 1995.

Yetman, Norman R. "Ex-Slave Interviews and the Historiography of Slavery." *American Quarterly* 36 (Summer 1984): 181–210.

———, ed. *Voices from Slavery: 100 Authentic Slave Narratives*. Mineola, N.Y.: Dover, 2000.

Index

Abdy, Edward Strutt, 84, 96, 177, 181
Abolitionism/abolitionists, 20, 60, 69, 147, 151, 178–79
Absentee plantation ownership, 50–51, 189
Adams, Lydia, 179
Africans: family structure in America, 155, 165, 197; importation of, in southern Louisiana, 32–33, in lowcountry South Carolina, 22–24, 50, 155, in northern Virginia, 14–15. *See also* Angola; Gambia; Gold Coast; Slave trade, transatlantic; Windward Coast
Agricultural work. *See* Labor, agricultural
Aime, Josephine, 135
Alabama, 176, 224n33, 231n23
Alderly plantation, 126. *See also* Ward, Joshua John
Alexander, John, 184. *See also* Litchfield plantation; Willbrook plantation
Alexander, William, 184. *See also* Litchfield plantation; Willbrook plantation
Alexandria, Virginia: domestic slave trade, 175–78, 180; economy, 17, 18, 92–93, 95; free black population, 93; slave hiring, 179–81; slaveholdings, 68, 122, 152; slaves, 69, 89, 91, 92–93, 178; during War of 1812, 18
Alexandria Gazette (Alexandria, Virginia), 17, 19, 21, 43, 120, 150, 174, 175, 176, 180–81, 183, 186
Allen, William, 186
Allison, Harrison, 182
Allston, Benjamin, 126, 157, 186, 188. *See also* Guendalos plantation
Allston, Charlotte A., 190
Allston, J. H., 185
Allston, John, 160

Allston, Robert F. W., 27, 46, 47, 48, 73, 102–3, 104, 123, 125, 126, 186–88, 189–90. *See also* Chicora Wood plantation; Guendalos plantation; Pipe Down plantation; Waverly plantation
Allston, William, 24
Alston, Benjamin, Jr., 154, 156, 184
Alston, Jacob Motte, 21, 46, 48–49, 50, 51, 75, 98, 102, 124–26, 128, 183, 184, 213n26. *See also* Woodbourne plantation
Alston, Thomas Pinckney, 186, 189, 190
Alston, William, 123, 184
Alston, William Algernon, Jr., 127. *See also* Calais plantation; Clifton plantation; Forlorn Hope plantation; Friendfield plantation; Marietta plantation; Michaux plantation; Rose Hill plantation (Georgetown District, South Carolina); Strawberry Hill plantation; Youngville plantation
American Revolution, 15, 24–25
American Union for the Relief and Improvement of the Colored Race, 178
Anderson, Richard O., 27–28
Andrews, Ethan Allen, 95, 120, 178, 181
Angola, 23
Antigua, 14
Arcenaux, Zenon, 195
Arkansas, 176
Arlington plantation, 18, 181. *See also* Custis, George W. P.; Lee, Robert E.
Armand, J. B., 132
Ascension Parish, Louisiana, 34, 60, 79, 108, 112, 168, 196
Ashby, Richard, 21

Bailey's Crossroads, Virginia, 122
Baltimore, Maryland, 177, 178
Bancroft, Frederic, 180
Banks, Henry, 70
Barbados, 14
Baton Rouge, Louisiana, 28, 34
Bayou Lafourche, Louisiana, 34, 169
Bayou Teche, Louisiana, 34
Beese, Welcome, 188
Bell, Frank, 39–40, 67, 70, 71, 152, 226n23
Bell, R., 82
Belmont plantation, 167, 169
Berlin, Ira, 13, 50, 58–59, 80, 111, 134, 140, 224n33
Berry, Daina Ramey, 10
Bertaua Brothers plantation, 193
Birdfield plantation, 104
Black, Maggie, 49
Black (river), 24, 128, 129, 155
Blackburn, Edward, 174
Blassingame, John W., 5–6, 9, 149, 208n7
Blouin, Daniel, 163–64
Boggess, Ann, 173
Boré, Etienne de, 30–31, 33
Boucry, Jean Baptiste, 197
Boundaries and opportunities, 4, 8–9, 11–12, 36, 39, 63, 87, 117, 141, 203–6; for slave family contact, 63–87; for slave family formation, 142–70; for slave family stability, 171–200; for slaves' internal economies, 88–113
Bowman, John, 186
Brazil, 29
Bremer, Frederika, 101–2, 127, 129
Brignac, Francois, 139
Broadwater, Charles Guy, 144
Brookgreen plantation, 50, 101, 126, 156. *See also* Ward, Joshua John
Brooks, Charlotte, 54, 80, 83
Brooks, Phillips, 181
Brown, Fred, 85, 165
Brown, Hagar, 77, 100
Brown, Henry, 75–76, 99
Bruce, Seddon, and Wilkins plantation, 105, 110
Brudot, Francis, 185
Bruin, Joseph, 175, 176, 177, 178–79
Bryant, Margaret, 75, 100, 102
Buckley, Joshua, 172
Buena Vista plantation, 131, 136, 164–65. *See also* Winchester, Benjamin

Buiew, Edward de, 78, 85
Bull Creek, South Carolina, 128
Bureau of Refugees, Freedmen, and Abandoned Lands, 164–65
Burke, James, 145
Burnside, John, 34, 78, 161, 165, 194, 198
Bush Hill plantation, 17, 19, 119, 147–48, 174, 182, 226n15. *See also* Scott, Richard Marshall, Jr.; Scott, Richard Marshall, Sr.; Scott, Virginia Gunnell

Cabahannocer plantation, 30
Calais plantation, 27. *See also* Alston, William Algernon, Jr.
Callouer, Jean Baptiste, 169
Camp, Stephanie, 146
Canada, 6, 45, 46, 65, 68, 93, 151, 173, 176, 179
Cantrelle, Michel, 163
Caribbean, the, 14, 25, 29, 30, 32, 51, 88, 123, 136
Carney, Judith, 50
Carolina, Albert, 53
Carter, Wormley, 174–75
Centreville, Virginia, 89, 122, 181
Champomier, P. A., 35 table 1.6, 139
Charleston, South Carolina: during American Revolution, 25; and domestic slave trade, 189; rice industry, 49, 124–25; and transatlantic slave trade, 23, 24
Chastant, Jean Baptiste, 193–94
Chenet, Pierre, 195
Chesapeake, the, 10, 22; during American Revolution, 15; grain production in, 16–17, 44–45; tobacco production in, 14–16, 40–41; and transatlantic slave trade, 14–15; during War of 1812, 18. *See also* Fairfax County, Virginia; Maryland; Virginia
Chicora Wood plantation, 27, 48, 50, 75, 76–77. *See also* Allston, Robert F. W.; Pringle, Elizabeth W. Allston
Childbearing/childrearing (among slaves), 65–70, 73–76, 81–84; breastfeeding, 65–66, 75–76, 79–80, 81; childbirth, 64–65, 73–74, 78–80, 188, 197, 217n30; in southern Louisiana, 77–86, 197, 217n30; pregnancy, 5, 64–65, 73, 78, 80, 182; in lowcountry South Carolina, 73–76, 188; supervision of children, 66–69, 74–75, 81–84; in northern Virginia, 64–70
Children (slave), 27, 63–70, 73–76, 78–87; of co-residential couples, 145, 149–52, 153–56,

161, 163–66, 174–75, 176, 183–91; of cross-plantation couples, 92, 144–45, 146–51, 152, 154, 159–60, 166–69, 172–83; and forced separation, 172–200; and internal economies, 100, 106, 110; in southern Louisiana, 34, 77–86, 106, 110, 134–37, 161, 163–69, 191–99; mortality rates, 65, 73–74, 80–81, 84, 137; orphans, 81, 150, 167, 195; of single parents, 145–46, 149–51, 166–68, 196; in lowcountry South Carolina, 27, 51, 73–77, 100, 126, 153–60, 183–91; in northern Virginia, 64–72, 92, 143–52, 172–83; work of, 34, 51, 67, 69–70, 76, 84–86
Christie, G.S.S., 189
Civil War, 21, 34, 147, 153, 183, 203
Clermont plantation, 19. *See also* Love, Charles J.
Clifton plantation, 127. *See also* Alston, William Algernon, Jr.
Clover Hill plantation, 147. *See also* Dulin, Edward
Coachman, W., 156, 159
Cockerill, Mastrom, 145
Coleman, George, 182
Coletrane, S., 189
Colomb, Octave, 86, 106, 107, 139, 198
Compton, John, 173–74
Conner, James, 60, 162–63
Constancia plantation. *See* Fagot, Samuel; Uncle Sam plantation
Cornelius, Catherine, 165
Cotton South, 21, 29, 80, 137, 198, 204, 217n29, 224n33; and domestic slave trade, 175–79, 188–89; and slave hiring, 55, 181, 231n23
Coulter, Peter, 172
Crane, Andrew, 134–35, 198–99
Craven, Avery Odell, 15
Croizet, Marie Elizabeth, 192
Custis, George W. P., 18. *See also* Arlington plantation

Davis, B. R., 151
Davis, John, 17, 146
Davis, Lizzie, 188–89
Demographic patterns: effects on family formation, 12, 92, 80–81, 142–70; in southern Louisiana, 10–11, 33–36, 79, 82, 130 table 5.3, 130–40, 160–70, 196–97, 224n35; sex ratios, 80–81, 122–23, 137, 139–40, 223n21; slave population growth, 14–15, 20 table 1.2, 22–23, 25, 27 table 1.4, 33, 34, 35 table 1.5, 35–36, 79, 82, 89, 118–19 table 5.1, 119, 120, 123, 124–25 table 5.2, 125–27, 129–30, 130 table 5.3, 131, 136–39, 175–76, 188, 196–97, 222–23n11, 224n35, 230n9, 232n37; in lowcountry South Carolina, 10, 22–23, 25, 27 table 1.4, 123–30, 124–25 table 5.2, 137, 152–60, 169–70, 188, 222–23n11, 223n21, 232n37; spatial distribution of slaves, 121–22, 127–29, 138–39; in northern Virginia, 10, 14–15, 16, 20 table 1.2, 89, 118–23, 118–29 table 5.1, 137, 142–52, 169–70, 175–76, 188, 230n9; white population growth, 16, 20 table 1.2, 27 table 1.4, 35 table 1.5. *See also* Slaveholding size
Derricks, Townshend, 147
Deyle, Steven, 175
Dicey, Edward, 89
Dipple plantation, 17. *See also* Scott, Richard Marshall, Sr.
Dirleton plantation, 127, 155, 157, 227n28. *See also* Sparkman, James Ritchie
Doar, David, 52, 158, 159
Doar, Stephen D., 186
Doby, Francis, 83, 109
Dorr, J. W., 36, 60
Dowling, Carter, 151–52
Drake, Pam, 60, 62, 82
Dranesville, Virginia, 122, 181
Du Bois, W. E. B., 5
Dufresne, Francois, 196
Dugas, Joseph, 195
Dulin, Edward, 147. *See also* Clover Hill plantation
Dunaway, Wilma, 7, 64, 142, 171
Duncan, Steven, 85, 168
Dunkin, Benjamin, 49
Duparc Brothers & Locoul plantation, 135, 161
Dusinberre, William, 73–74, 126, 189

Eber, Eboy, 194
Elkins, Stanley, 9
Emancipation, 3, 21, 27, 28, 54, 162, 164–65, 183, 203
Engerman, Stanley, 6–7, 9, 142, 171
Estate divisions, 12, 171, 199; in southern Louisiana, 191–93, 196–97; in lowcountry South Carolina, 154, 183–87; in northern Virginia, 119, 143–46, 149–52, 172–76, 180. *See also* Slave families, forced separation of

Europe, 15, 16, 18, 22, 30, 32, 54
Evans, Estwick, 135

Fabre, Joseph Laurent, 136, 192
Fabre, Joseph Paul, 192
Fabre, Leonard, 192
Fabre & Fabre plantation, 192. *See also* Croizet, Marie Elizabeth; Fabre, Joseph Laurent; Fabre, Joseph Paul; Fabre, Leonard
Fagot, Samuel, 132, 136, 164, 198. *See also* Uncle Sam plantation
Fairfax County, Virginia, 10, 14–21, 39–46, 63–72, 89–97, 118–23, 143–52, 172–83, 203–5; during American Revolution, 15; and domestic slave trade, 20, 119–20, 175–79; economic development in, 14–21, 119–21; emigration of whites from, 15–16, 19, 21, 176, 179; forced separation of slave families in, 172–83; geography of, 14, 121–22; Quakers in, 19–21; slave family formation in, 92, 143–52; slave hiring in, 20, 43, 89–91, 119, 120, 148, 177, 179–83, 199; slaveholding size in, 18–21, 44–45, 91, 118–21, 118–19 table 5.1, 143–52, 172, 172–83; slaves' internal economies in, 89–97; slaves' work in, 39–46, 69–72; slaves' childbearing/childrearing in, 63–70; and transatlantic slave trade, 14–15; during War of 1812, 17–18
Fairfax Court House, Virginia, 122, 181
Fairfield plantation, 155
Falls Church, Virginia, 122
Family life. *See* Slave families
Farmington, Virginia, 174
Fernand, Jean Baptiste, 198
Fitzhugh, John, 181
Fitzhugh, Maria, 151–52
Fitzhugh, M. C., 66
Fitzhugh, William, 149–51. *See also* Ravensworth plantation
Fletcher, Elijah, 16, 45, 65, 96, 143, 145–46
Fletcher, Rebecca, 60, 83, 84
Flowers, Annie, 168
Fogel, Robert W., 6–7, 9, 142, 171
Follett, Richard, 58, 61, 78, 80, 112, 140, 167, 217n29
Forlorn Hope plantation, 127. *See also* Alston, William Algernon, Jr.
France, 15, 16, 18, 28, 29, 30, 32, 54, 138
Franklin & Armfield, 175, 176, 178

Frans-Tracy, James, 182
Frazier, E. Franklin, 5
Free blacks: manumitted slaves, 20, 32, 91, 93, 120, 148, 172–73, 230n4; married to slaves, 91, 143, 147–48, 156, 167, 174, 182, 230n4; population of, in Fairfax County, Virginia, 20 table 1.2, in Georgetown District, South Carolina, 27 table 1.4, in St. James Parish, Louisiana, 35 table 1.5
Freedmen's Bureau. *See* Bureau of Refugees, Freedmen, and Abandoned Lands
French Revolution, 16–18
Friendfield plantation, 127, 157, 184–85, 227n28. *See also* Alston, William Algernon, Jr.; Sparkman, James Ritchie; Withers, Francis
Frobel, Anne, 43–44, 147. *See also* Wilton Hill plantation
Frobel, John J., 43–44. *See also* Wilton Hill plantation

Gambia, 23
Gang labor. *See* Labor, agricultural
Gaudet, Auguste, 193–94, 196
Genovese, Eugene D., 6, 9, 213n25
George, Ceceil, 60, 61, 79, 85
Georgetown District, South Carolina, 10, 21–28, 46–54, 72–77, 97–104, 123–30, 152–60, 183–91, 203, 205; absenteeism in, 50–51; during American Revolution, 24–25; and domestic slave trade, 188–90; economic development in, 21–28, 123–28; forced separation of slave families in, 183–91; geography of, 21–22, 24, 128–29; slave family formation in, 152–60; slave hiring in, 190–91; slaveholding size in, 10, 24, 26–27, 123–29, 124–25 table 5.2, 152–60, 183–91; slaves' internal economies in, 97–104; slaves' childbearing/childrearing in, 73–76, 188; slaves' work in, 46–54; and transatlantic slave trade, 22–24, 25
Georgia, 10, 25, 49, 52, 75, 77, 98–99, 102, 159, 176, 179, 189. *See also* Lowcountry, the
Glennie, Alexander, 128, 155
Goddard, Thomas F., 185
Godfrey, Ellen, 75
Gold Coast, 23
Golden Grove plantation, 133, 136, 193, 194. *See also* Shepherd, C. M.; Shepherd, J. H.; Shepherd, R. D.
Gordon & Forstall plantation, 33

Gore, Laura Locoul, 161
Gourdin, M., 198
Goutraux, J., 193
Grain cultivation. *See* Labor, agricultural
Great Britain, 15, 17, 18, 25, 29, 30, 85
Greenfield plantation, 129. *See also* Pringle, Julius Izard
Grey, David, 148, 182
Grigsby, Alexander, 119, 176
Guendalos plantation, 188. *See also* Allston, Benjamin; Allston, Robert F.W.
Guerin, François, 194
Gunnell, Robert, 173
Gutheim, Frederick, 14
Gutman, Herbert G., 6, 142

Hagley plantation, 127, 190. *See also* Weston, Plowden C.
Hall, Basil, 49, 52, 74–75, 99
Hamilton, Thomas, 61–62, 85–86, 137
Hamilton Station, Virginia, 152
Harned, William, 179
Henderson, Charles, 145
Henderson, Francis, 45–46, 66, 92, 151
Herbert, E., 198–99
Heriot, Edward Thomas, 127. *See also* Mount Arena plantation; Northampton plantation; Dirleton plantation
Heriot, E. J., 186
Hite, Elizabeth Ross, 61, 84–85, 109, 165–66, 197
Hodgson, Adam, 52, 99
Holland, 15
Hollin Hall plantation, 16, 45, 65, 145. *See also* Mason, Thomson
Homestead plantation, 34. *See also* Welham, W. P.
Horry, Ben, 28, 47, 50, 51, 53, 98, 101, 156, 157, 158
Horry, Peter, 25
Houmas plantation, 78, 79, 105, 165
Hubert, Henry, 182
Hudson, Larry E., Jr., 9, 92, 100, 103, 142
Hunter, Caroline, 66
Huntington, John, 176

Iberian peninsula, 18
Indigo, 22, 24, 25, 29–30
Ingraham, Joseph Holt, 85, 108, 132

Jackson, George, 45, 69–70, 90, 91, 93, 147
Jackson, Silas, 70, 90
Janney, John Jay, 42
Janney, Samuel, 18
Janney, Thomas, 182
Johnson, Daffney, 60, 84, 86, 166
Johnson, Octave, 110
Johnston, Dennis, 121
Jones, George, 168
Jordan, Daniel, 190
Joyner, Charles, 101

Keithfield plantation, 27–28. *See also* Anderson, Richard O.
Keller, Jean Baptiste, 195
Kenner, Alexander, 108
Kenner, Duncan, 112
Kentucky, 82, 176
King, William, 190
Kingsford, William, 110–11

Labor, agricultural: in family groups, 70–72, 76–77, 85–86; gang labor/time-work, 40–41, 44–46, 51, 59–61, 62, 77, 79, 86, 99, 205; in grain production, 16–21, 41–43, 44–46, 120–21; in southern Louisiana, 28–36, 54–62, 76–81, 86, 104–7, 130–37; in rice production, 21–27, 46–54, 73–74, 76–77, 123–27; sexual division of labor, 71–72, 76–77, 86–87; in lowcountry South Carolina, 21–27, 46–54, 73–74, 76–77, 123–27; in sugar production, 28–36, 54–62, 76–81, 84–86, 104–7, 130–37; task system/piece-work, 49–54, 59, 62, 72, 73, 75–77, 87, 97–100, 103, 104, 213n26; in tobacco production, 14–18, 29, 40–41, 44, 49–50, 62; in northern Virginia, 14–21, 29, 40–46, 49–50, 62, 71–72, 120–21
Labor, domestic: in southern Louisiana, 82, 85–86, 217n32; in lowcountry South Carolina, 52, 74, 127, 158, 184, 185; in northern Virginia, 43–44, 67–68, 69, 72
Labor, skilled: in southern Louisiana, 59, 61, 86, 105; in lowcountry South Carolina, 97–98; in northern Virginia, 89–90. *See also* Slave artisans
Lafourche Parish, Louisiana, 34, 169
Lance, Gabe, 53, 156
Landry, Jean Baptiste, 168
Lane, William, 176, 181

Lapice, P. M., 164, 167, 228n46
Lapice Brothers plantation, 131, 133
Laurel Hill plantation, 48–49, 190. *See also* Weston, Francis Marion; Weston, Plowden C.
LeBourgeois, Arnaud, 139
Lee, Francis Lightfoot, 176. *See also* Sully plantation
Lee, Robert E., 181, 231n23. *See also* Arlington plantation
Legg, E. P., 175
Leon, John A., 29, 32, 131, 132
Lewis, Eleanor Parke Custis, 138–39
Lewis, Lawrence, 138–39, 177, 178. *See also* Woodlawn plantation
Litchfield plantation, 127–28, 184. *See also* Tucker, John Hyrne
Longwood plantation, 126, 128. *See also* Ward, Joshua John
Loudoun County, Virginia, 42
Louisiana, 4, 10, 11, 13, 21, 28–36, 54–62, 77–87, 104–12, 130–40, 160–70, 176, 177, 189–90, 191–200, 203–4; colonial period, 28–32; and domestic slave trade, 32–33, 34, 137–38, 139–40, 189, 197–98; economic development, 28–36; geography of, 28–29, 31–32, 138–39; sugar production in, 28–36, 54–62, 76–81, 84–86, 104–7, 130–37; and transatlantic slave trade, 32–33; white migration to, 33–34
Love, Charles J., 19, 176. *See also* Clermont plantation
Love, Hunton, 84, 168, 196
Lowcountry, the, 4, 10, 21–28, 46–54, 72–77, 97–104, 123–30, 152–60, 183–91, 203, 205; absenteeism in, 50–51; during American Revolution, 24–25; colonization of, 21–23; economic development, 21–28, 123–28; geography of, 21–22, 24, 128–29; rice production in, 21–27, 46–54, 73–74, 76–77, 123–27; and transatlantic slave trade, 22–24, 25. *See also* Georgetown District, South Carolina; Georgia; South Carolina
Lower Mississippi Valley. *See* Louisiana
Lowndes, Charles, 190
Lowndes, William, 74, 101, 154, 156–57, 190
Lucas, Charles Peyton, 68
Lucerne plantation, 127. *See also* Read, John Harleston
Lyell, Charles, 20
Lynch, Thomas, 24

Madagascar, 22
Malaria, 23–24, 50, 73
Malet, William Wyndham, 99–100
Malone, Ann Patton, 9, 165, 166, 169, 192, 197
Manumission of slaves, 20, 32, 91, 93, 120, 148, 172–73, 230n4. *See also* Free blacks
Marietta plantation, 127. *See also* Alston, William Algernon, Jr.
Marion District, South Carolina, 188–89
Maroons, 25
Marriage (of slaves). *See* Slave families, family formation
Martin, Jonathan D., 180
Martineau, Harriet, 138
Maryland, 14, 131, 137, 178
Maryville plantation, 127. *See also* Read, John Harleston
Mason, Ann, 176
Mason, Barlow, 21, 176
Mason, Murray, 182
Mason, Thomson, 16, 90, 96. *See also* Hollin Hall plantation
Mather, George, 164, 193
McCarty, Daniel, the Younger, 15–16. *See also* Mount Air plantation
McCollam plantation, 168, 169
McDonald, John, 91
McDowell, Davison, 153–54, 156, 159
McInteer, Sarah, 176
McMillen, Sally, 64
Melancon, Joseph, 164, 193
Michaux plantation, 127. *See also* Alston, William Algernon, Jr.
Middleton, Henry Augustus, 183. *See also* Weehaw plantation
Mills, Robert, 128–29
Mingo Creek, South Carolina, 129
Minor, William, 168
Miscegenation: in southern Louisiana, 169; in lowcountry South Carolina, 157–58, 227n34; in northern Virginia, 148–49, 226n17
Mississippi (river), 28, 29, 31, 32, 34, 54, 111, 134, 135, 138, 139, 165, 169
Mississippi (state), 176, 181
Missouri, 176
Moody, V. Alton, 79
Morgan, Philip D., 9, 23, 44–45, 49–51, 101
Morin, Antoine, 31

Mount Air plantation, 15–16. *See also* McCarty, Daniel, the Younger
Mount Arena plantation, 127. *See also* Heriot, Edward Thomas
Mount Vernon plantation, 41, 90, 93, 118, 121, 149, 177. *See also* Washington, Bushrod; Washington, George
Mulattoes. *See* Miscegenation
Multiple plantation ownership, 17, 28, 68–69, 126–28, 153, 159, 183–84, 190, 192

Napoleonic wars, 18
Newman, William R., 181
New Orleans, Louisiana, 28, 34, 54, 138, 139; and domestic slave trade, 33, 134, 137, 140, 189–90, 197; economy, 29, 30; free black population, 32; and slave hiring, 32
Nichols, Christopher, 42, 45, 65, 68
Nichols, Heidi, 199
Nickens, Amos, 91
Northampton plantation, 127, 155, 157, 184–85, 227n28. *See also* Heriot, Edward Thomas; Withers, Francis
North Carolina, 50, 137, 188
Northup, Solomon, 55, 168

Oak Alley plantation, 194–95. *See also* Roman, Jacques Telesphore
Oakley plantation, 127. *See also* Read, John Harleston
Occoquan (city), Virginia, 122
Occoquan (river), 122
Offret, Alfred, 181
Ohio, 176
Oliphant, Laurence, 183–84
Oliver, William, 98, 99, 156
Olmsted, Frederick Law, 94, 97; in southern Louisiana, 31, 49, 61, 79–80, 81, 98–99, 108–9, 111–12, 132, 162, 197, 221n40; in the lowcountry, 51, 52, 75, 76, 77, 102, 104, 159, 213n25; in northern Virginia, 68, 92–93, 181
One, Margaret, 156
Orange Grove plantation, 34
Oryzantia plantation, 126. *See also* Ward, Joshua John
Overseers: in southern Louisiana, 60, 78–79, 165, 169, 198; in lowcountry South Carolina, 49, 52, 53, 76–77, 103, 158, 159; in northern Virginia, 41, 64, 71, 90, 96

Parish, Peter J., 7, 9, 204
Parker, Amos, 35, 60
Pass system, 90, 108, 146–47, 158–60, 168–69, 175, 185, 203
Patterson, Albert, 194
Payne, Oscar, 178
Pee Dee (river), 24, 27, 102, 125–29, 159
Pennsylvania, 173
Persac, A., 139, 169, 193
Petigru, Mary Ann, 186
Phillips, Ulrich Bonnell, 5, 7
Pinckney, Charles C., 156, 157
Pinckney, H. D., 160
Pipe Down plantation, 186, 187–88, 190. *See also* Allston, Robert F. W.
Plantation revolution: in southern Louisiana, 28–33; in lowcountry South Carolina, 21–24; in northern Virginia, 14–15
Pohoke plantation, 17, 95, 146
Pointe Coupee Parish, Louisiana, 32, 111
Polite, Sam, 158
Pollet, Louis, 195
Potomac (river), 14, 21, 91–92, 118, 121–22, 178, 181
Potter, James, 176
Potter, O., 185
Pregnancy, 5, 64; among slave women, in southern Louisiana, 78–79, 80, in lowcountry South Carolina, 73–74, in northern Virginia, 64–65, 182; among white women, 64. *See also* Childbearing/childrearing
Price, Birch & Co., 175, 176
Price, Thomas W., 186
Primogeniture, 119
Prince William County, Virginia, 93–94, 122
Pringle, Elizabeth W. Allston, 27, 46, 50, 75, 76, 129, 187–88. *See also* Chicora Wood plantation
Pringle, Julius Izard, 129, 155. *See also* White House plantation
Productivity: in grain cultivation, 18–20, 64, 94–95, 120–21; in rice cultivation, 24, 99, 124–27; in sugar cultivation, 33, 58, 107–8, 111, 131–36; in tobacco cultivation, 15–16
Proto-peasants (slaves as), 88–89, 96, 113
Pulszky, Francis & Theresa, 79–80, 84, 162, 165, 194
Punishment (of slaves): in southern Louisiana, 80, 162, 193, 196; in lowcountry South Carolina, 53–54, 77, 103, 189–90; in northern Virginia, 70–71, 177

Pyatt, John Francis, 28
Pyatt, Martha Allston, 28

Quakers, 19–21

Ratcliffe, Winifred, 181
Ravensworth plantation, 121, 147, 149–51. *See also* Fitzhugh, William
Read, John Harleston, 127. *See also* Lucerne plantation; Maryville plantation; Oakley plantation; Upton plantation
Redpath, James, 69, 120, 145, 147, 173, 178, 181
Reid, P. J., 173
Rice cultivation. *See* Labor, agricultural
Richfield plantation, 187
Richmond, Virginia, 165–66
Ripley, Eliza, 83, 105
Robinson, Solon, 60, 79, 90, 139
Rodrigue, John C., 133, 162
Rogers, George, 126, 128, 129
Roman, Alfred, 165
Roman, Andre B., 86, 107, 136
Roman, G. F., 162
Roman, Jacques, 131, 135, 136
Roman, Jacques Telesphore, 193, 194–95. *See also* Oak Alley plantation
Roman, Jean Jacques, 193
Roman, Sósthene, 139
Roman, Victorin, 192–93
Rose Hill plantation (Fairfax County, Virginia), 149
Rose Hill plantation (Georgetown District, South Carolina), 127, 186. *See also* Alston, William Algernon, Jr.
Ruffin, Edmund, 20, 120–21
Russell, Emily, 178–79
Russell, William Howard: in southern Louisiana, 61, 78–79, 83, 85, 86, 107, 108–9, 165, 169, 194; in lowcountry South Carolina, 27, 100, 101–2, 104
Rutledge, Sabe, 155–56

Saint-Domingue, 30, 31, 32
Sale/purchase of slaves: in southern Louisiana, 32–33, 34, 54, 83, 134, 137–38, 139–40, 193–98; in lowcountry South Carolina, 23–24, 25, 26, 125–26, 185–90; in northern Virginia, 14–15, 20, 120, 143, 145, 149–50, 172, 174–79, 180, 182. *See also* Slave families, forced separation of; Slave trade, domestic; Slave trade, transatlantic
Sampit (river), 24, 127, 128, 129, 186
Sandy Island, South Carolina, 156
Santee (river), 24, 128, 129
Savannah, Georgia, 25
Schexnaydre, Benjamin, 133–34
Schmitz, Mark, 133
Schwalm, Leslie A., 76, 99, 101, 103, 187
Scott, David Wilson, 41–43, 67, 146
Scott, Richard Marshall, Jr., 94–95, 99, 121, 147–48, 177, 180, 219n15, 226n15, 230n13. *See also* Bush Hill plantation
Scott, Richard Marshall, Sr., 17, 19, 68–69, 148, 180, 226n15. *See also* Bush Hill plantation; Dipple plantation
Scott, Sabret, 144–45
Scott, Virginia Gunnell, 119, 182, 226n15. *See also* Bush Hill plantation
Shackleford, John, 186
Shanks, Nelly, 174
Shekell, B. O., 177
Shepherd, C. M., 193. *See also* Golden Grove plantation
Shepherd, J. H., 136, 194. *See also* Golden Grove plantation
Shepherd, R. D., 136, 194. *See also* Golden Grove plantation
Singleton plantation, 129
Sitterson, J. Carlyle, 29, 133, 135, 169, 223n24
Slave agency, 3–9, 50–51, 142–43, 187–88, 204–6. *See also* Slave families; Slave resistance
Slave artisans: in southern Louisiana, 59, 61, 86, 105; in lowcountry South Carolina, 97–98, 155, 156, 160; in northern Virginia, 89–90. *See also* Labor, skilled
Slave drivers: in southern Louisiana, 59, 60, 162; in lowcountry South Carolina, 50–51, 53–54, 101; in northern Virginia, 71
Slave families, 3–12, 13–14, 36, 39, 62, 63, 87, 88–89, 112–13, 117–18, 140–41, 142–43, 169–70, 171, 199–200, 203–6; childbearing/childrearing, 65–70, 73–76, 81–84; family formation, 142–70; forced separation of, 171–200; internal economies, 88–113; in southern Louisiana, 77–87, 104–113, 160–70, 191–99, 203, 205; in lowcountry South

Carolina, 72–77, 87, 97–104, 113, 152–60, 170, 183–91, 199–200, 203, 205; structure of, 142–70; in northern Virginia, 63–72, 87, 89–97, 113, 143–52, 170, 172–83, 199–200, 203–5; and work, 63–87

Slave hiring: in southern Louisiana, 32, 55, 168, 198–99; in lowcountry South Carolina, 97–98, 190–91, 199; in northern Virginia, 20, 43, 89–91, 119, 120, 148, 177, 179–83, 199

Slaveholders. *See individuals by name*

Slaveholding size: development of, 10–11, 12, 18–21, 24, 26–27, 34, 44–45, 118–21, 118–19 table 5.1, 123–29, 124–25 table 5.2, 130–39, 130–31 table 5.3; effects on family formation, 12, 91, 117, 143–60, 160–69, 172; effects on family stability, 12, 117, 140–41, 171, 172–200; in southern Louisiana, 10–11, 34, 130–39, 130–31 table 5.3, 160–69, 191–99; in lowcountry South Carolina, 10, 24, 26–27, 123–29, 124–25 table 5.2, 152–60, 183–91; in northern Virginia, 18–21, 44–45, 91, 118–21, 118–19 table 5.1, 143–52, 172, 172–83

Slave resistance: escape from the South, 25, 45–46, 60, 65, 68, 69, 93, 147, 151, 155, 162–63, 176, 178–79, 181, 182; to hiring, 181–83, 198–99; in southern Louisiana, 32, 60, 110, 111, 162–63, 196, 198–99; rebellion, 32, 111; to sale, 178–79, 187–88, 196; in lowcountry South Carolina, 25, 51, 155, 187–88, 220n23; theft, 96–97; truancy, 110, 147, 178, 182–83; in northern Virginia, 45–46, 65, 68, 69, 93, 96–97, 147, 151, 176, 178, 179, 181–83; and work, 51, 71

Slave trade, domestic, 20, 25, 32–33, 120, 137–38, 139–40, 171, 175–80, 188–90, 197–98, 199

Slave trade, transatlantic, 14, 22–24, 25, 32–33

Smith, Charles, 144

Smith, James, 93

Smith, John W., 175

Snow, N., 191

Solomon, Daniel, 174

South Carolina, 4, 10, 21–28, 46–54, 72–77, 97–104, 123–30, 152–60, 183–91, 203, 205; absenteeism in, 50–51; during American Revolution, 24–25; colonization of, 21–23; economic development, 21–28, 123–28; geography of, 21–22, 24, 128–29; rice production in, 21–27, 46–54, 73–74, 76–77, 123–27; and transatlantic slave trade, 22–24, 25

Sparkman, James Ritchie, 52, 73, 76, 77, 102, 103, 190. *See also* Dirleton plantation; Friendfield plantation

Springfield plantation, 126. *See also* Ward, Joshua John

Stafford, Gracie, 165

Stampp, Kenneth M., 5, 9

Stevenson, Brenda E., 45, 64, 66, 69, 142, 225nn2,5

Steward, Austin, 45, 69, 70, 71, 90, 93–94, 149

St. Helena Island, South Carolina, 158

Still, William, 60

Stirling, James, 62

St. James Parish, Louisiana, 10–11, 14, 28–36, 54–62, 77–87, 104–13, 118, 130–41, 160–70, 191–99, 203, 205; and domestic slave trade, 34, 134, 137–38, 139–40, 193–98; economic development in, 34–36, 131–37; forced separation of slave families, 191–99; geography of, 28–29, 34–36, 138–39; slave family formation in, 160–70; slave hiring, 198–99; slaveholding size, 10–11, 34, 130–39, 130–31 table 5.3, 160–69, 191–99; slaves' internal production in, 104–113; slaves' childbearing/childrearing in, 77–87, 197, 217n30; slaves' work in, 54–62; and transatlantic slave trade, 32–33

Stone, Levy, 173

Strawberry Hill plantation, 127. *See also* Alston, William Algernon, Jr.

Stuart, James, 132

Sugar cultivation. *See* Labor, agricultural

Sully plantation, 149, 176. *See also* Lee, Francis Lightfoot

Sutcliffe, Robert, 92–93

Sweeney, George, 180

Sweig, Donald M., 121, 179, 179, 180

Tadman, Michael, 80, 82, 137, 139–40, 171, 175, 189

Task system. *See* Labor, agricultural

Taylor, John, 120

Tennessee, 19, 176

Thorpe, T. B., 35, 55, 56, 57, 106–8, 135

Tixier, Victor, 32, 36, 105, 111, 138, 168

Tobacco cultivation. *See* Labor, agricultural

Tower, Philo, 36, 54, 81
Triplett, George, 144
True Blue plantation, 74, 75, 76, 127, 155, 227n28. *See also* Weston, Plowden C.
Tucker, John Hyrne, 127–28, 184. *See also* Litchfield plantation; Willbrook plantation
Tureaud, Benjamin, 105, 106
Tyler, Elizabeth, 178

Uncle Sam plantation, 106, 132, 136, 164. *See also* Fagot, Samuel
Upton plantation, 127. *See also* Read, John Harleston
Uriell, Patrick, 195, 197

Veney, Bethany, 68
Vienna, Virginia, 39, 152, 226n23
Virginia, 10, 14–21, 39–46, 63–72, 89–97, 118–23, 143–52, 172–83, 203–5; during American Revolution, 15; and domestic slave trade, 20, 119–20, 175–79; economic development in, 14–21, 119–21; geography of, 14, 121–22; grain production in, 16–17, 44–45; tobacco production in, 14–16, 40–41; and transatlantic slave trade, 14–15; during War of 1812, 18

Waccamaw (river), 24, 125, 127, 128, 155, 156, 186, 189, 220n23, 227n28
Waccamaw Neck, South Carolina, 50, 101, 127–29, 157, 186–87, 189. *See also* Georgetown District, South Carolina
Wachesaw plantation, 156
Walney plantation, 45
Walsh, Lorena, 45
Ward, Joshua John, 28, 34, 50, 99, 101, 123, 126–27, 128, 156, 157, 184, 188, 220n24. *See also* Alderly plantation; Brookgreen plantation; Longwood plantation; Oryzantia plantation; Prospect Hill plantation; Springfield plantation
War of 1812, 17–18
Washington, Bushrod, 149, 177, 179. *See also* Mount Vernon plantation
Washington, D.C., 17, 93, 122, 147, 151, 174, 175, 177, 179
Washington, George, 16, 41, 93, 118, 172–73. *See also* Mount Vernon plantation

Waterford plantation, 127. *See also* Weston, Plowden C.
Watters, William, 173
Waverly plantation, 189. *See also* Allston, Robert F. W.
Webre, P., 164, 198
Weehawka plantation, 127. *See also* Weston, Plowden C.
Weehaw plantation, 155, 157. *See also* Middleton, Henry Augustus
Welham, W. P., 167, 169. *See also* Homestead plantation
West, Emily, 76, 142, 171
West, Matilda, 180
Weston, Emily Frances, 190
Weston, Francis Marion, 49. *See also* Laurel Hill plantation
Weston, Paul D., 154–55, 157, 186, 227n28
Weston, Plowden, 25
Weston, Plowden C., 51–52, 53, 73, 74, 76, 99, 104, 127, 158, 159, 184, 190. *See also* Hagley plantation; True Blue plantation; Waterford plantation; Weehawka plantation
Whatson, Cornelius, 178
Wheat cultivation. *See* Labor, agricultural
White House plantation, 129, 155. *See also* Pringle, Julius Izard
Wilkins, William Webb, 34–35, 198. *See also* Wilton plantation
Willbrook plantation, 128, 184. *See also* Tucker, John Hyrne
Williams, D. G., 191
Williams, George, 176
Williams, Harriet, 174
Willson, M., 191
Wilmington, North Carolina, 188
Wilson, Thomas, 191
Wilton Hill plantation, 43–44, 182. *See also* Frobel, John J.
Wilton plantation, 106, 107, 165, 198. *See also* Wilkins, William Webb
Winchester, Benjamin, 131, 133, 136, 161, 164. *See also* Buena Vista plantation
Windsor, Richard S., 175
Windward Coast, 23
Winyah Bay, South Carolina, 24, 28, 102, 104, 129, 155, 189

Winyah Intelligencer (Georgetown, South Carolina), 155, 185, 188
Withers, Francis, 184–85. *See also* Friendfield plantation; Northampton plantation
Wood, Peter, 22
Woodbourne plantation, 21, 48–49, 75, 98, 125, 128. *See also* Alston, Jacob Motte

Woodlawn plantation, 120, 121, 177. *See also* Lewis, Lawrence
Work patterns. *See* Labor, agricultural

Yellow fever, 23
Youngville plantation, 127. *See also* Alston, William Algernon, Jr.

NEW PERSPECTIVES ON THE HISTORY OF THE SOUTH

Edited by John David Smith

"In the Country of the Enemy": The Civil War Reports of a Massachusetts Corporal, edited by William C. Harris (1999)
The Wild East: A Biography of the Great Smoky Mountains, by Margaret L. Brown (2000; first paperback edition, 2001)
Crime, Sexual Violence, and Clemency: Florida's Pardon Board and Penal System in the Progressive Era, by Vivien M. L. Miller (2000)
The New South's New Frontier: A Social History of Economic Development in Southwestern North Carolina, by Stephen Wallace Taylor (2001)
Redefining the Color Line: Black Activism in Little Rock, Arkansas, 1940–1970, by John A. Kirk (2002)
The Southern Dream of a Caribbean Empire, 1854–1861, by Robert E. May (2002)
Forging a Common Bond: Labor and Environmental Activism during the BASF Lockout, by Timothy J. Minchin (2003)
Dixie's Daughters: The United Daughters of the Confederacy and the Preservation of Confederate Culture, by Karen L. Cox (2003)
The Other War of 1812: The Patriot War and the American Invasion of Spanish East Florida, by James G. Cusick (2003)
"Lives Full of Struggle and Triumph": Southern Women, Their Institutions, and Their Communities, edited by Bruce L. Clayton and John A. Salmond (2003)
German-Speaking Officers in the United States Colored Troops, 1863–1867, by Martin W. Öfele (2004)
Southern Struggles: The Southern Labor Movement and the Civil Rights Struggle, by John A. Salmond (2004)
Radio and the Struggle for Civil Rights in the South, by Brian Ward (2004; first paperback edition, 2006)
Luther P. Jackson and a Life for Civil Rights, by Michael Dennis (2004)
Southern Ladies, New Women: Race, Region, and Clubwomen in South Carolina, 1890–1930, by Joan Marie Johnson (2004)
Fighting Against the Odds: A Concise History of Southern Labor Since World War II, by Timothy J. Minchin (2004; first paperback edition, 2006)
"Don't Sleep With Stevens!": The J. P. Stevens Campaign and the Struggle to Organize the South, 1963–1980, by Timothy J. Minchin (2005)
"The Ticket to Freedom": The NAACP and the Struggle for Black Political Integration, by Manfred Berg (2005; first paperback edition, 2007)
"War Governor of the South": North Carolina's Zeb Vance in the Confederacy, by Joe A. Mobley (2005)
Planters' Progress: Modernizing Confederate Georgia, by Chad Morgan (2005)
The Officers of the CSS Shenandoah, by Angus Curry (2006)
The Rosenwald Schools of the American South, by Mary S. Hoffschwelle (2006)
Honor in Command: Lt. Freeman S. Bowley's Civil War Service in the 30th United States Colored Infantry, edited by Keith Wilson (2006)

A Black Congressman in the Age of Jim Crow: South Carolina's George Washington Murray,
   by John F. Marszalek (2006)
The Spirit and the Shotgun: Armed Resistance and the Struggle for Civil Rights,
   by Simon Wendt (2007; first paperback edition, 2010)
Making a New South: Race, Leadership, and Community after the Civil War,
   edited by Paul A. Cimbala and Barton C. Shaw (2007)
From Rights to Economics: The Ongoing Struggle for Black Equality in the U.S. South,
   by Timothy J. Minchin (2008)
Slavery on Trial: Race, Class, and Criminal Justice in Antebellum Richmond, Virginia,
   by James M. Campbell (2008; first paperback edition, 2010)
Welfare and Charity in the Antebellum South, by Timothy James Lockley (2008; first
   paperback edition, 2009)
T. Thomas Fortune the Afro-American Agitator: A Collection of Writings, 1880–1928,
   by Shawn Leigh Alexander (2008; first paperback edition, 2010)
Francis Butler Simkins: A Life, by James S. Humphreys (2008)
Black Manhood and Community Building in North Carolina, 1900–1930, by Angela
   Hornsby-Gutting (2009; first paperback edition, 2010)
Counterfeit Gentlemen: Manhood and Humor in the Old South, by John Mayfield
   (2009; first paperback edition, 2010)
*The Southern Mind Under Union Rule: The Diary of James Rumley, Beaufort, North
   Carolina, 1862–1865,* edited by Judkin Browning (2009, first paperback edition, 2011)
The Quarters and the Fields: Slave Families in the Non-Cotton South, by Damian Alan
   Pargas (2010; first paperback edition, 2011)
The Door of Hope: Republican Presidents and the First Southern Strategy, 1877–1933,
   by Edward O. Frantz (2011)
Painting Dixie Red: When, Where, Why, and How the South Became Republican,
   edited by Glenn Feldman (2011)
After Freedom Summer: How Race Realigned Mississippi Politics, 1965–1986,
   by Chris Danielson (2011)
*Dreams and Nightmares: Martin Luther King Jr., Malcolm X, and the Struggle for Black
   Equality in America,* by Britta Waldschmidt-Nelson (2012)
Hard Labor and Hard Time: Florida's "Sunshine Prison" and Chain Gangs,
   by Vivien M. L. Miller (2012)

www.ingramcontent.com/pod-product-compliance
Lightning Source LLC
Chambersburg PA
CBHW071833230426
43671CB00012B/1951